Lecture Notes in Computer Science 4890

Commenced Publication in 1973
Founding and Former Series Editors:
Gerhard Goos, Juris Hartmanis, and Jan van Leeuwen

Editorial Board

T0223135

Francesco Bonchi   Elena Ferrari
Bradley Malin   Yücel Saygin (Eds.)

Privacy, Security, and Trust in KDD

First ACM SIGKDD International Workshop, PinKDD 2007
San Jose, CA, USA, August 12, 2007
Revised Selected Papers

 Springer

Volume Editors

Francesco Bonchi
Pisa KDD Laboratory
ISTI - C.N.R, Pisa, Italy
E-mail: francesco.bonchi@isti.cnr.it

Elena Ferrari
University of Insubria
Department of Computer Science and Communication, Varese, Italy
E-mail: elena.ferrari@uninsubria.it

Bradley Malin
Vanderbilt University
Nashville, TN 37232, USA
E-mail: b.malin@vanderbilt.edu

Yücel Saygin
Sabanci University
Istanbul, Turkey
E-mail: ysaygin@sabanciuniv.edu

Library of Congress Control Number: 2008922886

CR Subject Classification (1998): H.4, H.3, C.2, H.2, D.4.6, K.4-6

LNCS Sublibrary: SL 4 – Security and Cryptology

ISSN   0302-9743
ISBN-10   3-540-78477-2 Springer Berlin Heidelberg New York
ISBN-13   978-3-540-78477-7 Springer Berlin Heidelberg New York

Springer is a part of Springer Science+Business Media

springer.com

© Springer-Verlag Berlin Heidelberg 2008

Typesetting: Camera-ready by author, data conversion by Scientific Publishing Services, Chennai, India
Printed on acid-free paper   SPIN: 12236466   06/3180   5 4 3 2 1 0

Preface

Vast amounts of data are collected by service providers and system administrators, and are available in public information systems. Data mining technologies provide an ideal framework to assist in analyzing such collections for computer security and surveillance-related endeavors. For instance, system administrators can apply data mining to summarize activity patterns in access logs so that potential malicious incidents can be further investigated. Beyond computer security, data mining technology supports intelligence gathering and summarization for homeland security. For years, and most recently fueled by events such as September 11, 2001, government agencies have focused on developing and applying data mining technologies to monitor terrorist behaviors in public and private data collections.

The application of data mining to person-specific data raises serious concerns regarding data confidentiality and citizens' privacy rights. These concerns have led to the adoption of various legislation and policy controls. In 2005, the European Union passed a data-retention directive that requires all telephone and Internet service providers to store data on their consumers for up to two years to assist in the prevention of terrorism and organized crime. Similar data-retention regulation proposals are under heated debate in the United States Congress. Yet, the debate often focuses on ethical or policy aspects of the problem, such that resolutions have polarized consequences; e.g., an organization can either share data for data mining purposes or it can not. Fortunately, computer scientists, and data mining researchers in particular, have recognized that technology can be constructed to support less polarized solutions. Computer scientists are developing technologies that enable data mining goals without sacrificing the privacy and security of the individuals to whom the data correspond.

To inject privacy into security and surveillance data mining projects, it is necessary to understand the goals of the latter. To further this exchange and highlight advances in research, we organized the First International Workshop on Privacy, Security, and Trust in KDD (PinKDD).

The First International Workshop on Privacy, Security, and Trust in KDD (PinKDD 2007) was held in conjunction with the 13th ACM SIGKDD International Conference on Knowledge Discovery and Data Mining. The workshop was held on August 12, 2007 in San Jose, California and brought together researchers, as well as practitioners, working on how privacy, security, and trust can be resolved or modeled within a data mining framework. The PinKDD workshop attracted considerable attention from the research community, as well as support from industrial organizations and academic institutions. The workshop received many high-quality research paper submissions, each of which was reviewed by a minimum of three members of the Program and Organizing Committee. In all, eight papers were selected for presentation at the workshop and inclusion in the workshop's post-proceedings. The papers represented the diversity of data mining research issues in privacy, security, and trust. In addition to two research sessions, the workshop highlights included a keynote talk which was delivered by Cynthia Dwork (Microsoft Research) and a spirited panel discussion on privacy issues in weblogs: the panel consisted of Ricardo Baeza-Yates (Yahoo Research), Cynthia Dwork (Microsoft Research), Lise Getoor (University of Maryland, College Park), and David Jensen (University of Massachusetts Amherst).

November 2007

Francesco Bonchi
Elena Ferrari
Bradley Malin
Yücel Saygin

Organization

Program Chairs

Francesco Bonchi
Pisa KDD Laboratory
ISTI - C.N.R.
Pisa, Italy
http://www-kdd.isti.cnr.it/~bonchi/

Elena Ferrari
University of Insubria
Department of Computer Science and Communication
Varese, Italy
http://www.dicom.uninsubria.it/~elena.ferrari/

Bradley Malin
Vanderbilt University
Department of Biomedical Informatics
Nashville, TN, USA
http://people.vanderbilt.edu/~b.malin

Yücel Saygin
Sabanci University
Faculty of Engineering and Natural Sciences
Istanbul, Turkey
http://people.sabanciuniv.edu/ysaygin/

Program Committee

Maurizio Atzori, ISTI-C.N.R., Italy
Roberto Bayardo, Google Inc., USA
Barbara Carminati, University of Insubria, Varese, Italy
Peter Christen, Australian National University, Canberra, Australia
Christopher Clifton, Purdue University, West Lafayette, USA
Josep Domingo-Ferrer, Rovira i Virgili University, Tarragona, Spain
Wenliang (Kevin) Du, Syracuse University, Syracuse, USA
Tyrone Grandison, IBM Almaden Research Center, USA
Satoshi Hada, IBM Tokyo Research Laboratory, Japan
Jutla Dawn, Saint Mary's University, Halifax, Canada
Murat Kantarcioglu, University of Texas, Dallas, USA
Hillol Kargupta, University of Maryland, Baltimore County, USA

Stan Matwin, University of Ottawa, Canada
Ilya Mironov, Microsoft Research, USA
Taneli Mielikäinen, Nokia Research Center, Palo Alto, USA
David Skillicorn, Queen's University, Ontario, Canada
Kian-Lee Tan, National University of Singapore
Bhavani Thuraisingham, University of Texas, Dallas, USA
Vicenç Torra, Spanish Scientific Research Council, Bellaterra, Spain
Vassilios Verykios, University of Thessaly, Volos, Greece
Ke Wang, Simon Fraser University, Canada
Jeffrey Yu, Chinese University of Hong Kong, S.A.R. China

Table of Contents

Invited Paper

Contributed Papers

An Ad Omnia Approach to Defining and Achieving Private Data Analysis

Cynthia Dwork

Microsoft Research
dwork@microsoft.com

Abstract. We briefly survey several privacy compromises in published datasets, some historical and some on paper. An inspection of these suggests that the problem lies with the nature of the privacy-motivated promises in question. These are typically syntactic, rather than semantic. They are also *ad hoc*, with insufficient argument that fulfilling these syntactic and *ad hoc* conditions yields anything like what most people would regard as privacy. We examine two comprehensive, or *ad omnia*, guarantees for privacy in statistical databases discussed in the literature, note that one is unachievable, and describe implementations of the other.

In this note we survey a body of work, developed over the past five years, addressing the problem known variously as statistical disclosure control, inference control, privacy-preserving datamining, and private data analysis. Our principal motivating scenario is a *statistical database*. A statistic is a quantity computed from a sample. Suppose a trusted and trustworthy curator gathers sensitive information from a large number of respondents (the sample), with the goal of learning (and releasing to the public) statistical facts about the underlying population. The problem is to release statistical information without compromising the privacy of the individual respondents. There are two settings: in the *non-interactive* setting the curator computes and publishes some statistics, and the data are not used further. Privacy concerns may affect the precise answers released by the curator, or even the set of statistics released. Note that since the data will never be used again the curator can destroy the data (and himself) once the statistics have been published.

In the *interactive* setting the curator sits between the users and the database. Queries posed by the users, and/or the responses to these queries, may be modified by the curator in order to protect the privacy of the respondents. The data cannot be destroyed, and the curator must remain present throughout the lifetime of the database.

There is a rich literature on this problem, principally from the satistics community [11, 15, 24, 25, 26, 34, 36, 23, 35] (see also the literature on controlled release of tabular data, contingency tables, and cell suppression), and from such diverse branches of computer science as algorithms, database theory, and cryptography [1, 10, 22, 28], [3, 4, 21, 29, 30, 37, 43], [7, 9, 12, 13, 14, 19, 8, 20]; see also the survey [2] for a summary of the field prior to 1989.

F. Bonchi et al. (Eds.): PinKDD 2007, LNCS 4890, pp. 1–13, 2008.

Clearly, if we are not interested in utility, then privacy can be trivially achieved: the curator can be silent, or can release only random noise. Throughout the discussion we will implicitly assume the statistical database has some non-trivial utility, and we will focus on the definition of privacy.

When defining privacy, or any other security goal, it is important to specify both what it means to compromise the goal and what power and other resources are available to the adversary. In the current context we refer to any information available to the adversary from sources other than the statistical database as *auxiliary information*. An attack that uses one database as auxiliary information to compromise privacy in a different database is frequently called a *linkage attack*. This type of attack is at the heart of the vast literature on hiding small cell counts in tabular data ("cell suppression").

1 Some Linkage Attacks

1.1 The Netflix Prize

Netflix recommends movies to its subscribers, and has offered a $1,000,000 prize for a 10% improvement in its recommendation system (we are not concerned here with how this is measured). To this end, Netflix has also published a training data set. According to the Netflix Prize rules webpage, "The training data set consists of more than 100 million ratings from over 480 thousand randomly-chosen, anonymous customers on nearly 18 thousand movie titles" and "The ratings are on a scale from 1 to 5 (integral) stars. *To protect customer privacy, all personal information identifying individual customers has been removed and all customer ids have been replaced by randomly-assigned ids.* The date of each rating and the title and year of release for each movie are provided" (emphasis added).

Netflix data are not the only movie ratings available on the web. There is also the International Movie Database (IMDb) site, where individuals may register for an account and rate movies. The users need not choose to be anonymous. Publicly visible material includes the user's movie ratings and comments, together with the dates of the ratings.

Narayanan and Shmatikov [32] cleverly used the IMDb in a linkage attack on the anonymization of the Netflix training data set. They found, "with 8 movie ratings (of which we allow 2 to be completely wrong) and dates that may have a 3-day error, 96% of Netflix subscribers whose records have been released can be uniquely identified in the dataset" and "for 89%, 2 ratings and dates are enough to reduce the set of plausible records to 8 out of almost 500,000, which can then be inspected by a human for further deanonymization." In other words, the removal of all "personal information" did not provide privacy to the users in the Netflix training data set. Indeed, Narayanan and Shmatikov were able to identify a particular user, about whom they drew several unsavory conclusions. Note that Narayanan and Shmatikov may have been correct in their conclusions or they may have been incorrect, but *either way this user is harmed.*

1.2 k-Anonymization and Sequelae

The most famous linkage attack was obtained by Sweeney [40], who identified the medical records of the governor of Massachusetts by linking voter registration records to "anonymized" Massachusetts Group Insurance Commission (GIC) medical encounter data, which retained the birthdate, sex, and zip code of the patient. Sweeney proposed an antidote: k-anonymity [38, 39, 41, 42]. Roughly speaking, this is a syntactic condition requiring that every "quasi-identifier" (essentially, combination of non-sensitive attributes) must appear at least k times in the published database, if it occurs at all. This can be achieved by coarsening attribute categories, for example, replacing 5-digit zipcodes by their 3-digit prefixes. There are many problems with k-anonymity (computational complexity and the fact that the choice of category coarsenings may reveal information about the database, to name two), but the biggest problem is that it simply does not provide strong privacy; a lot of information my still be leaked about respondents/individuals in the database. Machanavajjhala, Gehrke, and Kifer [30] discuss this problem, and respond by proposing a new criterion for the published database: ℓ-diversity. However, Xiao and Tao [43] note that multiple ℓ-diverse data releases completely compromise privacy. They propose a different syntactic condition: m-invariance.

The literature does not contain any direct attack on m-invariance (although, see Section 2.1 for general difficulties). However it is clear that something is going wrong: the "privacy" promises are syntactic conditions on the released datasets, but there is insufficient argument that the syntactic conditions have the correct semantic implications.

1.3 Anonymization of Social Networks

In a social network graph, nodes correspond to users (or e-mail accounts, or telephone numbers, etc), and edges have various social semantics (friendship, frequent communications, phone conversations, and so on). Companies that hold such graphs are frequently asked to release an anonymized version, in which node names are replaced by random strings, for study by social scientists. The intuition is that the anonymized graph reveals only the structure, not the potentially sensitive information of who is socially connected to whom. In [5] it is shown that anonymization does not protect this information at all; indeed it is vulnerable both to active and passive attacks. Again, anonymization is just an *ad hoc* syntactic condition, and has no privacy semantics.

2 On Defining Privacy for Statistical Databases

One source of difficulty in defining privacy for statistical databases is that the line between "inside"and "outside" is slightly blurred. In contrast, when Alice and her geographically remote colleague Bob converse, Alice and Bob are the "insiders," everyone else is an "outsider," and privacy can be obtained by any cryptosystem that is semantically secure against a passive eavesdropper.

Let us review this notion. Informally, semantic security says that the cipher-text (encryption of the message to be transmitted) reveals no information about the plaintext (the message). This was formalized by Goldwasser and Micali [27] along the following lines. The ability of the adversary, having access to both the ciphertext and any auxiliary information, to learn (anything about) the plaintext is compared to the ability of a party having access *only* to the auxiliary informa-tion (and not the ciphertext), to learn anything about the plaintext[1]. Clearly, if this difference is very, very tiny, then in a rigorous sense the ciphertext leaks (almost) no information about the plaintext.

The formalization of semantic security along these lines is one of the pillars of modern cryptography. It is therefore natural to ask whether a similar property can be achieved for statistical databases. However, unlike the eavesdropper on a conversation, the statistical database attacker is also a user, that is, a legitimate consumer of the information provided by the statistical database, so this attacker is both a little bit of an insider (not to mention that she may also be a respondent in the database), as well as an outsider, to whom certain fine-grained information should not be leaked.

2.1   Semantic Security for Statistical Databases?

In 1977 Tor Dalenius articulated an *ad omnia* privacy goal for statistical data-bases: anything that can be learned about a respondent from the statistical database should be learnable without access to the database. Happily, this for-malizes to semantic security (although Dalenius' goal predated the Goldwasser and Micali definition by five years). Unhappily, however, it cannot be achieved, both for small and big reasons. It is instructive to examine these in depth.

Many papers in the literature attempt to formalize Dalenius' goal (in some cases unknowingly) by requiring that the adversary's prior and posterior views about an individual (*i.e.*, before and after having access to the statistical database) shouldn't be "too different," or that access to the statistical database shouldn't change the adversary's views about any individual "too much." Of course, this is clearly silly, if the statistical database teaches us anything at all. For example, suppose the adversary's (incorrect) prior view is that everyone has 2 left feet. Access to the statistical database teaches that almost everyone has one left foot and one right foot. The adversary now has a very different view of whether or not any given respondent has two left feet. Even when used correctly, in a way that is decidedly not silly, this prior/posterior approach suffers from definitional awkwardness [21, 19, 8].

At a more fundamental level, a simple hybrid argument shows that it is im-possible to achieve cryptographically small levels of "tiny" difference between an adversary's ability to learn something about a respondent given access to the database, and the ability of someone without access to the database to learn something about a respondent. Intuitively, this is because the user/adversary is

---

[1] The definition in [27] deals with probabilistic polynomial time bounded parties. This is not central to the current work so we do not emphasize it in the discussion.

supposed to learn unpredictable and non-trivial facts about the data set (this is where we assume some degree of utility of the database), which translates to learning more than cryptographically tiny amounts about an individual. However, it may make sense to relax the definition of "tiny." Unfortunately, even this relaxed notion of semantic security for statistical databases cannot be achieved.

The final nail in the coffin of hope for Dalenius' goal is a formalization of the following difficulty. Suppose we have a statistical database that teaches average heights of population subgroups, and suppose further that it is infeasible to learn this information (perhaps for financial reasons) any other way (say, by conducting a new study). Consider the auxiliary information "Terry Gross is two inches shorter than the average Lithuanian woman." Access to the statistical database teaches Terry Gross' height. In contrast, someone without access to the database, knowing only the auxiliary information, learns much less about Terry Gross' height.

A rigorous impossibility result generalizes and formalizes this argument, extending to essentially any notion of privacy compromise. The heart of the attack uses extracted randomness from the statistical database as a one-time pad for conveying the privacy compromise to the adversary/user [16].

This brings us to an important observation: Terry Gross did not have to be a member of the database for the attack described above to be prosecuted against her. This suggests a new notion of privacy: minimize the increased risk to an individual incurred by joining (or leaving) the database. That is, we move from comparing an adversary's prior and posterior views of an individual to comparing the risk to an individual when included in, versus when not included in, the database. This new notion is called *differential privacy*.

Remark 1. It might be remarked that the counterexample of Terry Gross' height is contrived, and so it is not clear what it, or the general impossibility result in [16], mean. Of course, it is conceivable that counterexamples exist that would not appear contrived. More significantly, the result tells us that it is impossible to construct a privacy mechanism that both preserves utility and provably satisfies at least one natural formalization of Dalenius' goal. But proofs are important: they let us know exactly what guarantees are made, and they can be verified by non-experts. For these reasons it is extremely important to find *ad omnia* privacy goals and implementations that provably ensure satisfaction of these goals.

2.2 Differential Privacy

In the sequel, the randomized function $\mathcal{K}$ is the algorithm applied by the curator when releasing information. So the input is the data set, and the output is the released information, or *transcript*. We do not need to distinguish between the interactive and non-interactive settings.

Think of a database as a set of rows. We say databases D_1 and D_2 *differ in at most one element* if one is a subset of the other and the larger database contains just one additional row.

Definition 1. *A randomized function* $\mathcal{K}$ *gives* ϵ-differential privacy *if for all data sets* D_1 *and* D_2 *differing on at most one element, and all* $S \subseteq Range(\mathcal{K})$,

$$\Pr[\mathcal{K}(D_1) \in S] \leq \exp(\epsilon) \times \Pr[\mathcal{K}(D_2) \in S], \tag{1}$$

where the probability space in each case is over the coin flips of the mechanism $\mathcal{K}$.

A mechanism $\mathcal{K}$ satisfying this definition addresses all concerns that any participant might have about the leakage of her personal information: even if the participant removed her data from the data set, no outputs (and thus consequences of outputs) would become significantly more or less likely. For example, if the database were to be consulted by an insurance provider before deciding whether or not to insure a given individual, then the presence or absence of that individual's data in the database will not significantly affect her chance of receiving coverage.

Differential privacy is therefore an *ad omnia* guarantee. It is also a very strong guarantee, since it is a statistical property about the behavior of the mechanism and therefore is independent of the computational power and auxiliary information available to the adversary/user.

Differential privacy is not an absolute guarantee of privacy. As we have seen, any statistical database with any non-trivial utility can compromise privacy. However, in a society that has decided that the benefits of certain databases outweigh the costs, differential privacy ensures that only a limited amount of additional risk is incurred by participating in the (socially beneficial) databases.

Remark 2. 1. The parameter ϵ is public. The choice of ϵ is essentially a social question and is beyond the scope of this paper. That said, we tend to think of ϵ as, say, 0.01, 0.1, or in some cases, $\ln 2$ or $\ln 3$. If the probability that some bad event will occur is very small, it might be tolerable to increase it by such factors as 2 or 3, while if the probability is already felt to be close to unacceptable, then an increase of $e^{0.01} \approx 1.01$ might be tolerable, while an increase of e, or even only $e^{0.1}$, would be intolerable.
  2. Definition 1 extends to group privacy as well (and to the case in which an individual contributes more than a single row to the database). A collection of c participants might be concerned that their collective data might leak information, even when a single participant's does not. Using this definition, we can bound the dilation of any probability by at most $\exp(\epsilon c)$, which may be tolerable for small c. Of course, the point of the statistical database is to disclose aggregate information about large groups (while simultaneously protecting individuals), so we should expect privacy bounds to disintegrate with increasing group size.

3   Achieving Differential Privacy in Statistical Databases

We now describe an interactive mechanism, $\mathcal{K}$, due to Dwork, McSherry, Nissim, and Smith [20]. A *query* is a function mapping databases to (vectors of) real numbers. For example, the query "Count P" counts the number of rows in the database having property P.

When the query is a function f, and the database is X, the *true answer* is the value $f(X)$. The $\mathcal{K}$ mechanism adds appropriately chosen random noise to the true answer to produce what we call the *response*. The idea of preserving privacy by responding with a noisy version of the true answer is not new, but this approach is delicate. For example, if the noise is symmetric about the origin and the same question is asked many times, the responses may be averaged, cancelling out the noise[2]. We must take such factors into account.

Definition 2. *For $f : \mathcal{D} \to \mathbb{R}^d$, the* sensitivity *of f is*

$$\Delta f = \max_{D_1, D_2} \| f(D_1) - f(D_2) \|_1 \qquad (2)$$

for all D_1, D_2 differing in at most one element.

In particular, when $d = 1$ the sensitivity of f is the maximum difference in the values that the function f may take on a pair of databases that differ in only one element. For now, let us focus on the case $d = 1$.

For many types of queries Δf will be quite small. In particular, the simple counting queries discussed above ("How many rows have property P?") have $\Delta f = 1$. Our techniques work best – ie, introduce the least noise – when Δf is small. Note that sensitivity is a property of the function alone, and is independent of the database. The sensitivity essentially captures how great a difference (between the value of f on two databases differing in a single element) must be hidden by the additive noise generated by the curator.

On query function f the privacy mechanism $\mathcal{K}$ computes $f(X)$ and adds noise with a scaled symmetric exponential distribution with standard deviation $\sqrt{2}\Delta f/\epsilon$. In this distribution, denoted $\mathrm{Lap}(\Delta f/\epsilon)$, the mass at x is proportional to $\exp(-|x|(\epsilon/\Delta f))$.[3] Decreasing ϵ, a publicly known parameter, flattens out this curve, yielding larger expected noise magnitude. When ϵ is fixed, functions f with high sensitivity yield flatter curves, again yielding higher expected noise magnitudes.

The proof that $\mathcal{K}$ yields ϵ-differential privacy on the single query function f is straightforward. Consider any subset $S \subseteq Range(\mathcal{K})$, and let D_1, D_2 be any pair of databases differing in at most one element. When the database is D_1, the probability mass at any $r \in Range(\mathcal{K})$ is proportional to $\exp(-|f(D_1) - r|(\epsilon/\Delta f))$, and similarly when the database is D_2. Applying the triangle inequality in the

---

[2] We do not recommend having the curator record queries and their responses so that if a query is issued more than once the response can be replayed. One reason is that if the query language is sufficiently rich, then semantic equivalence of two syntactically different queries is undecidable. Even if the query language is not so rich, the devastating attacks demonstrated by Dinur and Nissim [14] pose completely random and unrelated queries.

[3] The probability density function of $\mathrm{Lap}(b)$ is $p(x|b) = \frac{1}{2b} \exp(-\frac{|x|}{b})$, and the variance is $2b^2$.

exponent we get a ratio of at most $\exp(-|f(D_1) - f(D_2)|(\epsilon/\Delta f))$. By definition of sensitivity, $|f(D_1) - f(D_2)| \leq \Delta f$, and so the ratio is bounded by $\exp(-\epsilon)$, yielding ϵ-differential privacy.

It is easy to see that for any (adaptively chosen) query sequence $f_1, \ldots, f_d$, ϵ-differential privacy can be achieved by running $\mathcal{K}$ with noise distribution $\mathrm{Lap}(\sum_i \Delta f_i/\epsilon)$ on *each* query. In other words, the quality of each answer deteriorates with the sum of the sensitivities of the queries. Interestingly, it is sometimes possible to do better than this. Roughly speaking, what matters is the maximum possible value of $\Delta = ||(f_1(D_1), f_2(D_1), \ldots, f_d(D_1)) - (f_1(D_2), f_2(D_2), \ldots, f_d(D_2))||_1$. The precise formulation of the statement requires some care, due to the potentially adaptive choice of queries. For a full treatment see [20]. We state the theorem here for the non-adaptive case, viewing the (fixed) sequence of queries $f_1, f_2, \ldots, f_d$ as a single d-ary query f and recalling Definition 2 for the case of arbitrary d.

Theorem 1. *For $f : \mathcal{D} \rightarrow \mathbb{R}^d$, the mechanism $\mathcal{K}_f$ that adds independently generated noise with distribution $\mathrm{Lap}(\Delta f/\epsilon)$ to each of the d output terms enjoys ϵ-differential privacy.*

Among the many applications of Theorem 1, of particular interest is the class of *histogram* queries. A histogram query is an arbitrary partitioning of the domain of database rows into disjoint "cells," and the true answer is the set of counts describing, for each cell, the number of database rows in this cell. Although a histogram query with d cells may be viewed as d individual counting queries, the addition or removal of a single database row can affect the entire d-tuple of counts in at most one location (the count corresponding to the cell to (from) which the row is added (deleted); moreover, the count of this cell is affected by at most 1, so by Definition 2, every histogram query has sensitivity 1.

4   Utility of $\mathcal{K}$ and Some Limitations

The mechanism $\mathcal{K}$ described above has excellent accuracy for insensitive queries. In particular, the noise needed to ensure differential privacy depends only on the sensitivity of the function and on the parameter ϵ. Both are independent of the database and the number of rows it contains. Thus, if the database is very large, the errors for many questions introduced by the differential privacy mechanism is relatively quite small.

We can think of $\mathcal{K}$ as a differential privacy-preserving interface between the analyst and the data. This suggests a line of research: finding algorithms that require few, insensentitive, queries for standard datamining tasks. As an example, see [8], which shows how to compute singular value decompositions, find the ID3 decision tree, carry out k-means clusterings, learn association rules, and learn anything learnable in the statistical queries learning model using only a relatively small number of counting queries. See also the more recent work on contingency tables (and OLAP cubes) [6].

It is also possible to combine techniques of secure function evaluation with the techniques described above, permitting a collection of data holders to cooperatively simulate $\mathcal{K}$; see [17] for details.

Recent Extensions. Sensitivity of a function f is a *global* notion: the worst case, over *all* pairs of databases differing in a single element, of the change in the value of f. Even for a function with high sensitivity, it may be the case that "frequently" – that is, for "many" databases or "much" of the time – the function is locally insensitive. That is, much of the time, adding or deleting a single database row may have little effect on the value of the function, even if the worst case difference is large.

Given any database D, we would like to generate noise according to the local sensitivity of f at D. Local sensitivity is itself a legitimate query ("What is the local sensitivity of the database with respect to the function f?"). If, for a fixed f, the local sensitivity varies wildly with the database, then to ensure differential privacy the local sensitivity must not be revealed too precisely. On the other hand, if the curator simply adds noise to $f(D)$ according to the local sensitivity of f at D, then a user may ask the query f several times in an attempt to guage the local sensitivity, which we have just argued cannot necessarily be safely learned with great accuracy. To prevent this, we need a way of *smoothing* the change in magnitude of noise used so that on locally insensitive instances that are sufficiently far from highly sensitive ones the noise is small. This is the subject of recent work of Nissim, Raskhodnikova, and Smith [33].

In some tasks, the addition of noise makes no sense. For example, the function f might map databases to strings, strategies, or trees. McSherry and Talwar address the problem of optimizing the output of such a function while preserving ϵ-differential privacy [31]. Assume the curator holds a database X and the goal is to produce an object y. In a nutshell, their *exponential mechanism* works as follows. There is assumed to be a *utility function* $u(X,y)$ that measures the quality of an output y, given that the database is X. For example, if the database holds the valuations that individuals assign a digital good during an auction, $u(X,y)$ might be the revenue, with these valuations, when the price is set to y. The McSherry-Talwar mechanism outputs y with probability proportional to $\exp(u(X,y)\epsilon)$ and ensures ϵ-differential privacy. Capturing the intuition, first suggested by Jason Hartline, that privacy seems to correspond to truthfulness, the McSherry and Talwar mechanism yields approximately-truthful auctions with nearly optimal selling price. Roughly speaking, this says that a participant cannot dramatically reduce the price he pays by lying about his valuation. Interestingly, McSherry and Talwar note that one can use the simple composition of differential privacy, much as was indicated in Remark 2 above for obtaining privacy for groups of c individuals, to obtain auctions in which no cooperating group of c agents can significantly increase their utility by submitting bids other than their true valuations.

Limitations. As we have seen, the magnitude of the noise generated by $\mathcal{K}$ increases with the number of questions. A line of research initiated by Dinur and Nissim indicates that this increase is inherent [14]. They showed that if the database is a vector x of n bits and the curator provides relatively accurate (within $o(\sqrt{n})$) answers to $n \log^2 n$ random subset sum queries, then by using linear programming the adversary can reconstruct a database x' agreeing with x in all but $o(n)$ entries, ie, satisfying $support(x - x') \in o(n)$. We call this *blatant non-privacy*. This result was later strengthened by Yekhanin, who showed that if the attacker asks the n Fourier queries (with entries ± 1; the true answer to query vector y is the inner product $\langle x, y \rangle$) and the noise is always $o(\sqrt{n})$, then the system is blatantly non-private [44].

Additional strengthenings of these results were obtained by Dwork, Mscherry, and Talwar [18]. They considered the case in which the curator can sometimes answer completely arbitrarily. When the queries are vectors of standard normals and again the true answer is the inner product of the database and the query vector, they found a sharp threshold $\rho^* \approx 0.239$ so that if the curator replies completely arbitrarily on a $\rho < \rho*$ fraction of the queries, but is confined to $o(\sqrt{n})$ error on the remaining queries, then again the system is blatantly non-private even against only $O(n)$ queries. Similar, but slightly less strong results are obtained for ± 1 query vectors.

These are not just interesting mathematical exercises. While at first blush simplistic, the Dinur-Nissim setting is in fact sufficiently rich to capture many natural questions. For example, the rows of the database may be quite complex, but the adversary/user may know enough information about an individual in the database to uniquely identify his row. In this case the goal is to prevent any single *additional* bit of information to be learned from the database. (In fact, careful use of hash functions can handle the "row-naming problem" even if the adversary does not know enough to uniquely identify individuals at the time of the attack, possibly at the cost of a modest increase in the number of queries.) Thus we can imagine a scenario in which an adversary reconstructs a close approximation to the database, where each row is identified with a set of hash values, and a "secret bit" is learned for many rows. At a later time the adversary may learn enough about an individual in the database to deduce sufficiently many of the hash values of her record to identify the row corresponding to the individual, and so obtain her "secret bit." Thus, naming a set of rows to specify a query is not just a theoretical possibility, and the assumption of only a single sensitive attribute per user still yields meaningful results.

Research statisticians like to "look at the data." Indeed, conversations with experts in this field frequently involve pleas for a "noisy table" that will permit highly accurate answers to be derived for computations that are not specified at the outset. For these people the implications of the Dinur-Nissim results are particularly significant: no "noisy table" can provide very accurate answers to too many questions; otherwise the table could be used to simulate the interactive mechanism, and a Dinur-Nissim style attack could be mounted against the table. Even worse, while in the interactive setting the noise can be adapted to the queries, in the non-interactive setting the curator does not have this freedom to aid in protecting privacy.

5 Conclusions and Open Questions

We have surveyed a body of work addressing the problem known variously as statistical disclosure control, privacy-preserving datamining, and private data analysis. The concept of ϵ-differential privacy was motivated and defined, and a specific technique for achieving ϵ-differential privacy was described. This last involves calibrating the noise added to the true answers according to the sensitivity of the query sequence and to a publicly chosen parameter ϵ.

Of course, statistical databases are a very small part of the overall problem of defining and ensuring privacy. How can we sensibly address privacy in settings in which the boundary between "inside" and "outside" is completely porous, for example, in outsourcing of confidential data for processing, bug reporting, and managing cookies? What is the right notion of privacy in a social network (and what are the questions of interest in the study of such networks)?

We believe the notion of differential privacy may be helpful in approaching these problems.

References

[1] Achugbue, J.O., Chin, F.Y.: The Effectiveness of Output Modification by Rounding for Protection of Statistical Databases. INFOR 17(3), 209–218 (1979)

[2] Adam, N.R., Wortmann, J.C.: Security-Control Methods for Statistical Databases: A Comparative Study. ACM Computing Surveys 21(4), 515–556 (1989)

[3] Agrawal, D., Aggarwal, C.C.: On the design and Quantification of Privacy Preserving Data Mining Algorithms. In: Proceedings of the 20th Symposium on Principles of Database Systems, pp. 247–255 (2001)

[4] Agrawal, R., Srikant, R.: Privacy-Preserving Data Mining. In: Proceedings of the ACM SIGMOD International Conference on Management of Data, pp. 439–450. ACM Press, New York (2000)

[5] Backstrom, L., Dwork, C., Kleinberg, J.: Wherefore Art Thou r3579x?: Anonymized Social Networks, Hidden Patterns, and Structural Steganography. In: Proceedings of the 16th International World Wide Web Conference, pp. 181–190 (2007)

[6] Barak, B., Chaudhuri, K., Dwork, C., Kale, S., McSherry, F., Talwar, K.: Privacy, Accuracy, and Consistency Too: A Holistic Solution to Contingency Table Release. In: Proceedings of the 26th Symposium on Principles of Database Systems, pp. 273–282 (2007)

[7] Beck, L.L.: A Security Mechanism for Statistical Databases. ACM TODS 5(3), 316–338 (1980)

[8] Blum, A., Dwork, C., McSherry, F., Nissim, K.: Practical Privacy: The SuLQ framework. In: Proceedings of the 24th ACM SIGMOD-SIGACT-SIGART Symposium on Principles of Database Systems (June 2005)

[9] Chawla, S., Dwork, C., McSherry, F., Smith, A., Wee, H.: Toward Privacy in Public Databases. In: Proceedings of the 2nd Theory of Cryptography Conference (2005)

[10] Chin, F.Y., Ozsoyoglu, G.: Auditing and infrence control in statistical databases, IEEE Trans. Softw. Eng. SE-8(6), 113–139 (April 1982)

[11] Dalenius, T.: Towards a Methodology for Statistical Disclosure Control. Statistik Tidskrift 15, 429–222 (1977)
[12] Denning, D.E.: Secure Statistical Databases with Random Sample Queries. ACM Transactions on Database Systems 5(3), 291–315 (1980)
[13] Denning, D., Denning, P., Schwartz, M.: The Tracker: A Threat to Statistical Database Security. ACM Transactions on Database Systems 4(1), 76–96 (1979)
[14] Dinur, I., Nissim, K.: Revealing Information While Preserving Privacy. In: Proceedings of the Twenty-Second ACM SIGACT-SIGMOD-SIGART Symposium on Principles of Database Systems, pp. 202–210 (2003)
[15] Duncan, G.: Confidentiality and statistical disclosure limitation. In: Smelser, N., Baltes, P. (eds.) International Encyclopedia of the Social and Behavioral Sciences, Elsevier, New York (2001)
[16] Dwork, C.: Differential Privacy. In: Bugliesi, M., Preneel, B., Sassone, V., Wegener, I. (eds.) ICALP 2006. LNCS, vol. 4052, pp. 1–12. Springer, Heidelberg (2006)
[17] Dwork, C., et al.: Our Data, Ourselves: Privacy Via Distributed Noise Generation. In: Vaudenay, S. (ed.) EUROCRYPT 2006. LNCS, vol. 4004, pp. 486–503. Springer, Heidelberg (2006)
[18] Dwork, C., McSherry, F., Talwar, K.: The Price of Privacy and the Limits of LP Decoding. In: Proceedings of the 39th ACM Symposium on Theory of Computing, pp. 85–94 (2007)
[19] Dwork, C., Nissim, K.: Privacy-Preserving Datamining on Vertically Partitioned Databases. In: Franklin, M. (ed.) CRYPTO 2004. LNCS, vol. 3152, pp. 528–544. Springer, Heidelberg (2004)
[20] Dwork, C., McSherry, F., Nissim, K., Smith, A.: Calibrating Noise to Sensitivity in Private Data Analysis. In: Proceedings of the 3rd Theory of Cryptography Conference, pp. 265–284 (2006)
[21] Evfimievski, A.V., Gehrke, J., Srikant, R.: Limiting Privacy Breaches in Privacy Preserving Data Mining. In: Proceedings of the Twenty-Second ACM SIGACT-SIGMOD-SIGART Symposium on Principles of Database Systems, pp. 211–222 (2003)
[22] Dobkin, D., Jones, A., Lipton, R.: Secure Databases: Protection Against User Influence. ACM TODS 4(1), 97–106 (1979)
[23] Fellegi, I.: On the question of statistical confidentiality. Journal of the American Statistical Association 67, 7–18 (1972)
[24] Fienberg, S.: Confidentiality and Data Protection Through Disclosure Limitation: Evolving Principles and Technical Advances, IAOS Conference on Statistics, Development and Human Rights (September 2000),
http://www.statistik.admin.ch/about/international/
fienberg_final_paper.doc
[25] Fienberg, S., Makov, U., Steele, R.: Disclosure Limitation and Related Methods for Categorical Data. Journal of Official Statistics 14, 485–502 (1998)
[26] Franconi, L., Merola, G.: Implementing Statistical Disclosure Control for Aggregated Data Released Via Remote Access, Working Paper No. 30, United Nations Statistical Commission and European Commission, joint ECE/EUROSTAT work session on statistical data confidentiality (April 2003),
http://www.unece.org/stats/documents/2003/04/confidentiality/
wp.30.e.pdf
[27] Goldwasser, S., Micali, S.: Probabilistic Encryption. J. Comput. Syst. Sci. 28(2), 270–299 (1984)
[28] Gusfield, D.: A Graph Theoretic Approach to Statistical Data Security. SIAM J. Comput. 17(3), 552–571 (1988)

[29] Lefons, E., Silvestri, A., Tangorra, F.: An analytic approach to statistical databases. In: 9th Int. Conf. Very Large Data Bases, pp. 260–274. Morgan Kaufmann, San Francisco (1983)

[30] Machanavajjhala, A., Gehrke, J., Kifer, D., Venkitasubramaniam, M.: l-Diversity: Privacy Beyond k-Anonymity. In: Proceedings of the 22nd International Conference on Data Engineering (ICDE 2006), p. 24 (2006)

[31] McSherry, F., Talwar, K.: Mechanism Design via Differential Privacy. In: Proceedings of the 48th Annual Symposium on Foundations of Computer Science (2007)

[32] Narayanan, A., Shmatikov, V.: How to Break Anonymity of the Netflix Prize Dataset. How to Break Anonymity of the Netflix Prize Dataset, http://www.cs.utexas.edu/~shmat/shmat_netflix-prelim.pdf

[33] Nissim, K., Raskhodnikova, S., Smith, A.: Smooth Sensitivity and Sampling in Private Data Analysis. In: Proceedings of the 39th ACM Symposium on Theory of Computing, pp. 75–84 (2007)

[34] Raghunathan, T.E., Reiter, J.P., Rubin, D.B.: Multiple Imputation for Statistical Disclosure Limitation. Journal of Official Statistics 19(1), 1–16 (2003)

[35] Reiss, S.: Practical Data Swapping: The First Steps. ACM Transactions on Database Systems 9(1), 20–37 (1984)

[36] Rubin, D.B.: Discussion: Statistical Disclosure Limitation. Journal of Official Statistics 9(2), 461–469 (1993)

[37] Shoshani, A.: Statistical databases: Characteristics, problems and some solutions. In: Proceedings of the 8th International Conference on Very Large Data Bases (VLDB 1982), pp. 208–222 (1982)

[38] Samarati, P., Sweeney, L.: Protecting Privacy when Disclosing Information: k-Anonymity and its Enforcement Through Generalization and Specialization, Technical Report SRI-CSL-98-04, SRI Intl. (1998)

[39] Samarati, P., Sweeney, L.: Generalizing Data to Provide Anonymity when Disclosing Information (Abstract). In: Proceedings of the Seventeenth ACM SIGACT-SIGMOD-SIGART Symposium on Principles of Database Systems, p. 188 (1998)

[40] Sweeney, L.: Weaving Technology and Policy Together to Maintain Confidentiality. J. Law Med. Ethics 25(2-3), 98–110 (1997)

[41] Sweeney, L.: k-anonymity: A Model for Protecting Privacy. International Journal on Uncertainty, Fuzziness and Knowledge-based Systems 10(5), 557–570 (2002)

[42] Sweeney, L.: Achieving k-Anonymity Privacy Protection Using Generalization and Suppression. International Journal on Uncertainty, Fuzziness and Knowledge-based Systems 10(5), 571–588 (2002)

[43] Xiao, X., Tao, Y.: M-invariance: Towards privacy preserving re-publication of dynamic datasets. In: SIGMOD 2007, pp. 689–700 (2007)

[44] Yekhanin, S.: Private communication (2006)

Phoenix: Privacy Preserving Biclustering on Horizontally Partitioned Data

Waseem Ahmad and Ashfaq Khokhar

Department of Electrical and Computer Engineering
University of Illinois at Chicago
1020 SEO, 851 South Morgan Street
M/C 154 Chicago, IL 60607-7124, USA
{wahmad,ashfaq}@ece.uic.edu

Abstract. Emerging business models require organizations to collaborate with each other. This collaboration is usually in the form of distributed clustering to find optimal customer targets for effective marketing. This process is hampered by two problems (1) Inability of traditional clustering algorithm in finding local (subspace) patterns in distributed data and (2) Privacy policies of individual organizations limiting the process of information sharing. In this paper, we propose an efficient privacy preserving biclustering algorithm on horizontally partitioned data, referred to as Phoenix, which solves both of these problems. It assumes a malicious adversary model which is more practical than commonly employed semi-honest adversary model. It is shown to outperform traditional K-means clustering algorithm in identifying local patterns. The distributed secure implementation of the algorithm is evaluated to be very efficient both in computation and communication requirements.

1   Introduction

Emerging business models require organizations to collaborate with each other in order to meet customer requirements effectively. These collaborations are mostly in the form of Distributed Data Clustering (DDC). DDC is used to find optimal customer targets for effective marketing and to build customer profiles for customized services. Information sharing among organizations improves the accuracy of cluster models. For example it will be hard for an organization to build effective customer profiles based just on the customer data available at its own sites. However, if organizations in the same market decide to share their customer data then relatively accurate customer profiles can be built. For instance online retailers may share their shoppers' information with each other so that customized recommendations can be served whenever a shopper moves from one site to the other.

Data partitioning for collaborative data analysis has been studied from two angles (1) Horizontal Partitioning where each site has the same attributes(features)

F. Bonchi et al. (Eds.): PinKDD 2007, LNCS 4890, pp. 14–32, 2008.

but different objects and (2) Vertical Partitioning where each site has same objects but their attributes are distributed across all sites. In this paper, we will limit ourselves to horizontal partitioning only.

Despite above mentioned benefits, clustering on data partitioned across administrative boundaries suffers from following drawbacks.

1. Clustering techniques are only good at finding global patterns. Note that clustering is achieved by maximizing the similarity within a class and minimizing the similarity across classes. The similarity criterion is based on some distance function computed over all attributes. This similarity comparison based on the entire set(space) of attributes tends to overlook local patterns where different objects are similar based on only a subset(subspace) of attributes. For example, two customers who have similar music taste but different taste in food and clothes would most likely be treated as belonging to different clusters under traditional clustering techniques. The fact that these customers have similar music taste is a local pattern which can't be identified by these techniques.
2. The second challenge is that information sharing across administrative boundaries poses a serious threat to individuals' privacy. In order to satisfy customer requirements and to comply with government regulations, most organizations have to devise privacy policies which strongly prohibit information sharing with other organizations. Under these policies, organizations share their raw data under Non-disclosure agreements and clustering is performed in a centralized warehouse model. Like other researchers [1, 2], we believe centralized warehouse model is limited in scope because of its inherent lack of scalability and associated costs. Therefore distributed clustering solutions are required which could also provide sufficient guarantees for the preservation of the privacy of the data.

Contributions

This paper is aimed at solving the above mentioned problems for those applications which require identification of local patterns from horizontally partitioned data in a privacy preserving manner. Our proposed solution, which is referred to as Phoenix, makes following contributions to this end.

Local Pattern Discovery through Biclustering. Identification of local (subspace) patterns is achieved through Biclustering. Biclustering is a technique which is capable of finding local patterns where a subset of objects(records) might be similar to each other based on only a subset of attributes. Biclusters and clusters are illustrated in Figure 1 for comparison. As is shown in the figure, biclusters can just cover part of rows or columns and may overlap with each other. We employ a bigraph crossing minimization based biclustering algorithm, namely cHawk, which was proposed in [3]. The input data is viewed as adjacency matrix and barycenter heuristic [4] is employed for crossing minimization. This algorithm is not only capable of finding local patterns with high accuracy and

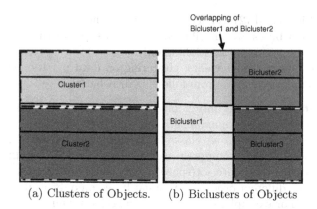

(a) Clusters of Objects.   (b) Biclusters of Objects

Fig. 1. Illustration of Clusters and Biclusters for Comparative Analysis

efficiency, it is also easily amenable to bandwidth and communication efficient secure implementation for privacy preserving biclustering.

Secure Aggregation in Malicious Adversary Model. We show that the proposed crossing minimization based Biclustering algorithm requires only a secure aggregation protocol and a secure Euclidean Distance function for a privacy preserving implementation. We provide these protocols under Secure Multiparty Computation setting. The secure aggregation protocol is based on the framework that we proposed in [5]. The salient features of this protocol are (1) Threshold Additively homomorphic cryptosystem with distributed key generation which eliminates the requirement of a trusted dealer, (2) Zero knowledge proofs to determine if the protocol was executed honestly and (3) a scalable hierarchical communication framework which is resilient to network failures. The proposed secure aggregation protocol has good communication complexity and requires only $O(\log \mathcal{N})$ rounds to provide a stable solution where $\mathcal{N}$ is the number of participants. This protocol is constructed for a malicious adversary model which assumes that participants may not follow the protocol honestly. Moreover protocol also caters for collusion whereby multiple malicious participants may collaborate against honest users. The only restriction is that number of colluding participants should not exceed the honest ones.

Many approaches have been proposed recently to provide privacy preserving distributed *clustering* solutions [6,7,8]. Even though *biclustering* has been shown to be very useful in fine-grained pattern discovery [3,9], there has been surprisingly little focus on providing distributed privacy preserving solutions to this problem. To our knowledge, privacy preserving distributed biclustering is being discussed for the first time in this paper.

Rest of the paper is organized as follows. Related research work is presented in Section 2. Proposed approach for privacy preserving biclustering, Phoenix, is presented in Section 3. Complexity analysis of Phoenix is carried out in Section 4. Experimental evaluation is presented in Section 5 and conclusions are drawn in Section 6.

2   Related Work

Privacy issues in statistical databases have been discussed comprehensively in [10, 11]. Recently privacy preserving data mining has become a very active area of research. Early solutions were proposed for privacy preserving decision tree construction [12, 13]. Privacy preserving solutions for association rule mining were proposed in [14, 15]. We will limit our focus on privacy preserving clustering solutions along with generic biclustering algorithms.

2.1   Privacy Preserving Clustering

There have been several solutions proposed for privacy preserving clustering problem. The main theme behind such solutions is the processing of the private data such that privacy is not compromised while accuracy of the cluster models remains as high as possible. There are two threads of research in this regard i.e. (1) Statistical Processing based approaches [16, 17] and (2) Cryptographic Secure Multiparty Computations(SMC) based approaches [15, 2, 8].

Statistical Processing Approaches can be categorized into model based clustering and clustering over randomized data.

1. Model based approaches [16] for privacy sensitive distributed clustering involve building cluster models at local sites and then transmitting them to a central server. A global model is then built which is an aggregate function of the local models.
2. Random Perturbation schemes [17] try to preserve user data by adding random noise to it while attempting to make sure that the necessary statistical aggregates such as mean don't get disturbed much. These protocols involve a privacy accuracy tradeoff and are not suitable for distributed privacy preserving clustering.

Secure Multiparty Computation (SMC) based approaches [2, 15, 8] employ cryptographic protocols which provide guarantees that each party would not learn more than the aggregate cluster models and its own personal data. SMC protocols essentially allow computations over encrypted data so that initial data sets and intermediate results are hidden from all participants. Only when the computations are complete, participants engage in a distributed protocol to decrypt the aggregated model.

Privacy preserving $k-$means clustering solutions were proposed in vertically partitioned data in [2] and in horizontally partitioned data in [7]. Moreover, a privacy preserving clustering solution based on EM (*Expectation Maximization*) mixture models was proposed in [6]. These solutions were proposed under the assumption of a semi-honest adversary model whereby participants are assumed to follow the protocol honestly. This assumption is not valid in many real life settings among mutually un-trusted participants. Moreover, the secure aggregation framework of [6] is based on a ring based synchronous communication framework which is unsuitable for large scale collaborations. The secure permutation algorithm of [2] is required to be executed for every item in the data set which makes it unscalable for large data sets. Another problem with these approaches, as mentioned above, is that clustering algorithms are unsuitable for discovery of local patterns which is the subject of this paper.

2.2 Biclustering

Biclustering has been shown to effectively discover local(sub-space) patterns in biological data sets [9,3], in text mining [18] and in collaborative filtering [19,20].

In our previous work, referred to as cHawk [3], we presented a biclustering algorithm which was based on bigraph crossing minimization. Salient features of the algorithm are as follows.

- We provided a theoretical connection between spectral partitioning and crossing minimization of a bipartite graph. Using this connection, a Biclustering Model based on Crossing minimization was proposed.
- An efficient implementation of the proposed model, termed as cHawk, was provided. The input data is viewed as adjacency matrix and barycenter heuristic [4] is used for solving the crossing minimization problem efficiently. Convergence of this heuristic was theoretically and experimentally proved in [4, 21]. We note that crossing minimization reorders the vertices on both layers of the bipartite graph such that vertices belonging to the same bicluster are brought into the vicinity of each other. This essentially reduces the bicluster identification problem from a global search to local search. Asymptotic complexity of barycenter heuristic is only $O(|E| + |V|log|V|)$ where E is the set of edges and V is the set of vertices in the input graph.
- An efficient algorithm for bicluster identification was proposed. This algorithm employs local search and is capable of finding constant, coherent and overlapped biclusters amid noise. The underlying similarity test is based on Bregman Divergence [22]. Bregman Divergence is defined as follows [22].

  If f is a strictly convex real-valued function, the f-entropy of a discrete measure $p(x) \geq 0$ is defined by $H_f(p) = -\sum_x (f(p(x)))$ and the Bregman divergence $B_f(p; q)$ is given as

$$B_f(p; q) = -\sum_x f(p(x)) - f(q(x)) - \nabla f(q(x))(p(x) - q(x)) \qquad (1)$$

When $f(x) = x \log x$, H_f is the Shannon entropy and $B_f(p; q)$ is the I-divergence, when $f(x) = -\log(x)$ we get the Burg entropy and discrete Itakura-Saito distortion

$$B_f(p; q) = \sum_x (\log \frac{q(x)}{f(x)} + \frac{p(x)}{q(x)} - 1)$$

Detailed comparison of cHawk with other biclustering approaches along with its implementation and evaluation is presented in [3].

3   Privacy Preserving Biclustering

Another advantage of crossing minimization based biclustering approach, apart from its efficiency, is that it is easily amenable to distributed secure implementations. For distributed implementation of the algorithm, we need to develop distributed versions of crossing minimization and bicluster identification algorithms. As we show in the coming sections, the distributed implementations are communication efficient which lead to efficient secure implementations as well.

3.1   Model for Distributed Biclustering

We assume that data is horizontally partitioned across a set $S : \{s_1 \ldots s_N\}$ of N Servers such that each server s_i has the same set $A : \{a_1 \ldots a_L\}$ of L attributes and a set $R_i : \{r_{i1} \ldots r_{iM_i}\}$ of M_i different objects(records) such that global object set R can be represented as $R = \bigcup_{\forall i \in S} R_i$.

Distributed Crossing Minimization. Each server can build bigraph on its local data such that objects are represented by nodes on one layer (say $Layer0$ nodes) and the attributes are represented by nodes on the other layer ($Layer1$). It would then perform crossing minimization on the Bigraph. The crossing minimization procedure requires iterative computation of the ranks of object vertices and the attribute vertices which are represented by two layers of the bigraph. This procedure starts by initializing ranks of vertices in one layer to random values. The rank for each vertex on the other layer of the bigraph is then calculated as a function of ranks of its neighbors on the first layer. In case of barycenter heuristic [3], this function is a weighted mean (μ) of the ranks of neighbors. The procedure terminates when there is no further change in the vertex ranks.

Let v_j represent the j'th object node in $Layer_0$ and set N_j represent the set of attribute nodes connected to v_j. Also let r_i represent the rank of i'th member of the set N_j. Then the weighted mean μ_j which represents the rank of j'th object node is given in Equation 2.

$$\mu_j = \frac{\sum_{i \in N_j} w_{i,j} \times r_i}{\sum_{i \in N_j} w_{i,j}} \tag{2}$$

Since we assume horizontal partitioning, each object node in the bigraph built over local data will have all necessary information (i.e. Ranks of its neighbor attribute nodes) to calculate its rank locally using Equation 2. This implies that no inter-server communication is required for calculating the ranks of objects at each server. On the other hand since attribute nodes are connected to object nodes which might be separated across different servers, we will have to engage in inter-server communication for exact computation of the ranks of these attribute nodes. Each server i shares its local value of μ_{ij} for j'th attribute to all other servers and then global weighted mean is calculated. This global weighted mean $\mu_G^{(j)}$ for an attribute $a_j \in A$ is simply the mean of all μ_{ij} values over n servers.

$$\mu_G^{(j)} = \frac{\sum_{1 \le i \le n} \mu_{ij}}{n}$$

The global weighted mean for an attribute node is the same over all servers and thus results in assignments of the same unique rank to each attribute node over all servers.

Distributed Bicluster Identification. By the end of the above mentioned distributed crossing minimization algorithm, each server will have a reordered representation of its local data. Since ranks of object nodes are calculated locally, each server does not know the global rank of its object nodes. Global rank of object nodes is useful only if each server knows the ranks of objects on other servers too. This is because of the fact that bicluster identification procedure works on contiguously ranked object nodes. It will have to be implemented in a distributed setting by having a hash table so that each server s_i can lookup the table to determine the set of servers S_{lookup} which have object nodes adjacent to its own object node. The server s_i would then engage in distributed bicluster identification procedure in collaboration with those servers which belong to the set S_{lookup}. There are two problems with this approach which are listed below.

1. It would result in tremendous increase in communication overhead. Firstly, all servers will have to engage in all-to-all broadcast of local ranks of each object node so as to assign unique global rank to each object node at each server. This is also required for building the required look up table. Given the large number of object nodes, the communication cost for this process would be prohibitively high. Secondly, during the bicluster identification process each server will have to engage in distributed bicluster identification with other servers belonging to the set S_{lookup}. This would require bicluster identification procedure to be performed in a synchronized manner at each server. Given the fact that these servers will be communicating on internet speeds and are distributed geographically, synchronization requirement will be impractical for most practical applications.

2. If each server knows the global rank of objects on other servers, it would amount to serious privacy breaches during the bicluster identification stage. Since crossing minimization step assigns similar global ranks to similar objects, knowing the global rank of objects on other servers is equivalent to knowing their similarity with local object nodes. This in itself reveals the attribute values of those objects as these values ought to be similar to the values of local objects with similar global ranks.

The solution to above mentioned problems is to employ a simple approach whereby object ranks are computed locally at each server and then bicluster identification process is performed locally as well. Once each server has identified its local biclusters, it broadcasts representatives of these biclusters to all other servers. A bicluster representative is a vector consisting of the mean of each attribute in the bicluster. Upon receiving the bicluster representatives from other servers, each server determines through Euclidean distance if its local biclusters can be combined with those from other servers. In case it finds strong similarity between one of its biclusters and incoming bicluster representatives. It updates the local bicluster representative for that bicluster such that the attribute means now reflect the global attribute means for the bicluster.

3.2   Model for Distributed Privacy Preserving Biclustering

Before we discuss the privacy issues with the above mentioned distributed implementation of the crossing minimization based biclustering approach, we first outline the assumed adversarial model.

Adversarial Model. We assume that the adversary can corrupt up to $t \leq \frac{n}{2}$ servers and corrupted servers can collude as well. The computational power of the adversary is modeled by a probabilistic polynomial time Turing machine. Moreover, the adversary is assumed to be static i.e. it chooses the corrupted servers at the beginning of the protocol. We also assume that the servers under adversarial control can be malicious i.e. they may not follow the protocol properly. This malicious adversary model is more realistic in real life applications as opposed to semi-honest model where adversary is supposed to follow the protocol.

Privacy Issues with Distributed Biclustering. We discussed distributed implementation of crossing minimization based biclustering approach in a previous section. Here we outline the privacy issues with this distributed implementation. These privacy issues are listed as follows.

1. The distributed implementation of the crossing minimization procedure requires each server to share its local rank for each attribute with other servers. Since rank of each attribute node is computed as a weighted mean of the ranks of the neighbor object nodes, it leaks information regarding edge weights which are actual matrix values.

   Given the fact that each server has the same rank for each attribute and all servers share the same set of Attributes $\mathcal{A}$, knowing the weighted mean of attribute ranks enables the adversary to guess about edge weights used in calculating the mean. For example, if an object node has a weighted mean μ. An adversary can attack the edge weights as follows. Lets assume there are two attribute nodes with ranks r_1 and r_2. This information is available to adversary. The only unknown information is edge weights x_1 and x_2. So according to Equation 2.

$$\frac{r_1 \times x_1 + r_2 \times x_2}{x_1 + x_2} = \mu$$

Now if r_1 happens to be 0, then we are left with

$$\frac{r_2 \times x_2}{x_1 + x_2} = \mu$$

Since r_2 is known to adversary and the edge weights are usually taken from a finite discrete space, the adversary can determine values of x_1 and x_2 efficiently even by a brute force attack.

2. Another problem emerges when bicluster representatives are broadcasted to all other servers for possible merging of the bicluster. Bicluster representatives are only means of attribute values in the bicluster. For small sized biclusters, mean of the attribute value is exposed to privacy attack mentioned above.

In the wake of above mentioned problems, we need to provide secure primitives using which all servers can calculate the global means for attribute nodes securely. Also required is a way to merge distributed biclusters over different servers securely.

Closer inspection of the first problem reveals that we need a mechanism by which all servers can engage in a secure aggregation protocol which makes sure that no server learns more than the aggregate global mean value and its own value at the end of the protocol. Clearly this can be achieved through secure multiparty computation protocols. We proposed a scalable secure aggregation protocol in [5] which is based on a threshold homomorphic cryptosystem described in Section 3.3. It has all desired properties to cater for adversarial model stated above. Salient features of this homomorphic cryptosystem are as follows.

1. Homomorphic property to ensure secure multiparty computation (See Section 3.3 for details).
2. Distributed key Generation to eliminate the requirement of trusted dealer.
3. Threshold decryption to cater for the colluding nodes.
4. Use of the zero knowledge proofs to determine if a specific server has completely followed the protocol or not.

For the problem of determining bicluster similarity securely, we allow broadcast of bicluster representatives in an encrypted form. Secure Euclidean distance protocol of [23] is then used to allow merging of similar biclusters.

In the next section we outline the construction of above mentioned threshold homomorphic cryptosystem. We will then use this cryptosystem to perform secure aggregation of the μ values for calculating the global mean thus realizing the crossing minimization securely. The secure bicluster identification and merging procedure will be detailed in Section 3.6.

3.3   Threshold Additively Homomorphic Cryptosystem

Salient features of the proposed additively homomorphic cryptosystem were described in the previous section. An additively homomorphic cryptosystem has the nice property that for two plain text messages m_1 and m_2, it holds $E(m_1)*E(m_2)= E(m_1 + m_2)$. This essentially means that we can have the sum of two numbers without knowing what those numbers were. The concept of threshold cryptography allows us to distribute shares of the private key among the set of servers such that until t of them collaborate, ciphertext can't be decrypted.

Construction of different features of the cryptosystem is outlined below.

Distributed Key Generation. Let there be $\mathcal{N}$ servers $(S_1, \ldots, S_{\mathcal{N}})$. We employ Distributed Key Generation protocol of Malkin et al [24] in our implementation. At the end of computation, an RSA modulus $n = pq$ is publicly known. All servers involved in the computation are convinced that the modulus is a product of two primes but no one knows the factors of n. Moreover each server is left with a private key share d_i generated using t-out-of-$\mathcal{N}$ sharing protocol of [24]. For verification of the decryption protocol, we need the following fixed public values: v which generates the cyclic group of squares in $\mathbb{Z}^*_{n^2}$ and for each decryption server a verification key $v_i = v^{\Delta d_i} \bmod (n^2)$ where $\Delta = \mathcal{N}!$.

Encryption. Let m be a message to be encrypted where $m \in \mathbb{Z}_{n^2}$. A random number r where $r \in \mathbb{Z}_n^*$ is selected. Ciphertext c is computed as: $c = (n+1)^m r^{n^2} \bmod n^2$.

Share Decryption. The i'th server will compute $c_i = c^{2\Delta d_i}$, Along with this will be a zero knowledge proof as that $\log_{c^4}(c_i^2) = \log_v(v_i)$ which will convince other servers that it has indeed raised c to his secret exponent d_i.

Share Combining. If each server has number of verified shares $\geq t+1$, then it can combine them into the result by following Jurik et al's scheme of [25] which combines subset S of shares of honest nodes as follows.

$$c' = \prod_{i \in S} c_i^{2\lambda_{0,i}^S}$$

where

$$\lambda_{0,i}^S = \Delta \prod_{j \in S_1} \frac{-j}{i-j} \in Z$$

The message m can be obtained from c' by applying the discrete log algorithm of [25].

Homomorphic Property of the Cryptosystem. To show that the cryptosystem is additively homomorphic, consider two messages $m1$ and $m2$ which are encrypted using the same public key pk such that $c_1 = E_{(pk)}(m1, r1)$ and $c_2 = E_{(pk)}(m2, r2)$ then $c_1 c_2 = g^{m_1} g^{m_2} r_1^{n^2} r_2^{n^2} = g^{m_1+m_2} r^{n^2}$ where $r = (r_1 r_2) \in \mathbb{Z}_n^*$ so $c_1 c_2 = E_{(pk)}(m1 + m2, r)$.

3.4   Scalable Communication Framework

Our secure aggregation framework [5] employs the self-stabilizing hierarchical communication algorithm of [26]. The algorithm assumes unique identifiers given to each Server. The Server with the minimum identifier plays the role of the root in the spanning tree. In this algorithm nodes(servers) which are part of some spanning tree expect to receive "power" from the root of the tree where "power" refers to continuous flow of certain messages one per round. The basic idea is that only legal roots may be the source of the power and the fake roots are forced out to make a new tree. Whenever a node receives power from a node with smaller identifier (than the one it is currently attached to), it attaches itself to that node's tree. In an asynchronous network, the power supply idea is implemented using different types of messages. Nodes use periodic exchange of "weak messages" to synchronize their state while "strong" messages are used to carry power. This algorithm is known to stabilize in $O(\mathcal{N})$ rounds without any knowledge of $\mathcal{N}$ (total number of servers).

3.5   Secure Crossing Minimization

Now that we have defined the primitive to perform secure aggregation, we can revisit the crossing minimization procedure to implement it securely. As mentioned previously, secure implementation of the crossing minimization only requires the secure aggregation of the local values of the weighted means for each attribute so as to calculate global mean for that attribute. This global mean is then used to assign a unique rank to each attribute. Note that we don't perform global calculation of mean in case of object nodes for reasons discussed in previous section. The secure distributed crossing minimization procedure is described in Algorithm 1.

As shown in Algorithm 1, the secure aggregation is performed in Step 7 through Step 14. Basically each server encrypts its local weighted mean for the attribute with public key PK using encryption function of Section 3.3 and broadcasts to all other servers. Each server combines these encrypted values using procedure of Section 3.3. Each server then uses its share s_i of the secret key to partially decrypt the encrypted value of the global key. These partial decryptions are then broadcasted to all servers. Each server verifies these partial decryptions using zero knowledge proof of Section 3.3 to determine if the protocol was honestly followed. If there are more than t honest servers, each server can combine their shares to get the decrypted value of global mean. Note that zero knowledge proof only makes sure that the protocol was honestly followed i.e. the server used its local share s_i to perform partial decryption. It does not secure against wrong values of the local means which are sent according to the protocol. To protect against such attacks, Wagner's resilient aggregation protocol [27] can be employed.

3.6   Secure Merging of Biclusters

Once the crossing minimization step is performed, each server performs bicluster identification procedure [3]. A set of biclusters are thus obtained which don't represent the global biclusters. Note that there is no predefined restriction on the number of biclusters. The bicluster identification procedure identifies each bicluster with dissimilarity score less than δ and number of rows and columns more than their respective threshold values.

To obtain the global bicluster model, each server builds a bicluster representative of each of its local biclusters. The bicluster representative is just a

---

Algorithm 1. Secure Distributed Crossing Minimization

Require: Bigraph BG, Public Key PK, Share of the private key S_i
Ensure: An embedding of BG (new ordering of the nodes in the two layers) which results in minimal number of crossings for BG.

1: $positionChanged \Leftarrow 1$
2: $DynamicLayer \Leftarrow 1$
3: **while** $PositionChanged \neq 0$ **do**
4:     $PositionChanged = 0$
5:     **for all** i such that $Node_i$ belongs to the nodes in the Dynamic Layer **do**
6:         Compute Weighted mean for $Node_i$ using Equation 2
7:         **if** $DynamicLayer$ represents the attribute nodes **then**
8:             Use encryption function of the cryptosystem from Section 3.3 to encrypt local values of weight mean
9:             Broadcast encrypted values of the weighted mean to all other servers
10:            Perform the Homomorphic combine operation of Section 3.3 to obtain an ecnrypted version of the global mean.
11:            Use Share Decryption protocol of Section 3.3 to determine the shares of the honest nodes.
12:            Use share combine protocol of Section 3.3 to combine shares of the honest nodes to achieve decrypted global mean value μ_G for the attribute node
13:            $WeightedMean = \mu_G$
14:         **end if**
15:         **if** $Node_i.Rank \neq WeightedMean$ **then**
16:            $Node_i.Rank = WeightedMean$
17:            $PositionChanged = PositionChanged + 1$
18:         **end if**
19:     **end for**
20:     Sort all the nodes in Dynamic Layer
21:     Now adjust node ranks such that each node has a unique rank
22:     $DynamicLayer = (1 - DynamicLayer)$
23: **end while**

---

mean of each attribute value in the bicluster over all objects which are members of the bicluster. Since biclusters are merged using a similarity score calculated through Euclidean distance, we follow the scheme of [23] to compute it securely. According to the scheme, the mean value v_{ai} of each attribute a_i is represented as $v_{ai}^2, -2 \times v_{ai}, -1$. This representation is broadcasted to all servers. Each server represents mean m_{ai} of each attribute a_i in its local bicluster as $1, m_{ai}, m_{ai}^2$. The dot product of these components will then be $v_{ai}^2 - 2 \times v_{ai} \times m_{ai} - m_{ai}^2 = (v_{ai} - m_{ai})^2$. We can extend this idea to the entire set $A : \{a_1 \ldots a_l\}$ of attributes as follows.

$$\sum_i (v_{ai} - m_{ai})^2 =$$

$$(\sum_i v_{ai}^2, -2 \times v_{a1}, \ldots, v_{al}, 1).(1, m_{a1}, \ldots, m_{al}, -\sum_i m_{ai}^2)$$

The only challenge now is to compute this dot product securely. For this purpose we can use the secure scalar product protocol of [23]. Each server determines the similarity of each of its local biclusters with received bicluster representatives from other servers by virtue of above mentioned secure Euclidean distance function. If two biclusters have similarity score more than a threshold value β, these biclusters are merged by virtue of an update of their representatives. The new bicluster representative will have values which are means of the values of the two similar biclusters. Thus we calculate the global bicluster representative securely.

4   Complexity Analysis of Phoenix

4.1   Computation Complexity

Key generation protocol would require selecting random numbers and performing primality tests till required conditions are met. In a naive implementation, $O(n^2)$ probes might be required till a suitable value of n (RSA modulus) is found. Practical considerations for efficient implementation of distributed key generation process are outlined in [24]. Their proposed method of distributed sieving considerably improves the efficiency of the key generation process.

Lets assume that $\mid \mathcal{R} \mid = \mathcal{N}\bar{\mathcal{R}}$ denotes the total number of rows (objects) of the input data matrix where $\mathcal{N}$ is the total number of servers and $\bar{\mathcal{R}}$ is the average number of rows at each server. Also $\mid \mathcal{A} \mid = m$ total number of columns of the input matrix and $\bar{\mathcal{R}}_{ci}$ = average number of rows per bicluster at server S_i. Also $\bar{\mathcal{A}}_{ci}$ = average number of columns per bicluster and k_i = number of biclusters at server S_i and I = average number of iterations of crossing minimization process. Now $O(\bar{\mathcal{R}})$ = time to compute weighted means at each server, $O(m)$ = time to perform encryption and decryption of attribute ranks during every second iteration and $O(\bar{\mathcal{R}} \log \bar{\mathcal{R}})$ = time to perform sorting based on means and $O(\bar{\mathcal{R}})$ = time taken in adjusting node positions.

Also $O(k_i \bar{\mathcal{A}}_{ci} \bar{\mathcal{R}}_{ci})$ = time to identify biclusters (when there is no overlap). In case of overlapped biclusters, the bicluster identification process would take $O(dk_i \bar{\mathcal{A}}_{ci} \bar{\mathcal{R}}_{ci})$ where d is the average degree of overlap among biclusters. Computation of bicluster representatives and encrypting them at Server S_i would require time in $O(\bar{\mathcal{R}}_{ci} \bar{\mathcal{A}}_{ci})$. Comparing the received b bicluster representatives with local biclusters would yield computation cost of $O(bk_i)$.

Given the fact that number of iterations are usually very small (I was never more than 5 in our experimental evaluation), the computation cost is usually bounded by $O(dk_i \bar{\mathcal{A}}_{ci} \bar{\mathcal{R}}_{ci})$ which is linear with the size of the local matrix as long as degree of overlap d remains constant.

4.2   Communication Overhead

Using the scalable communication framework, distributed Key generation process requires $O(P \log \mathcal{N})$ communication rounds where $\mathcal{N}$ is the total number of servers and P is the total number of probes.

The distributed crossing minimization process is very communication efficient. Messages are exchanged only while updating the ranks of attribute nodes. $O(Im \log \mathcal{N})$ messages rounds are required during crossing minimization where length of each message is equivalent to the number of bits of RSA modulus ($512 - 1024$ bits for most applications). $O(m \log \mathcal{N})$ message rounds would be required for share combining and decryption over all iterations of the crossing minimization process. It should be noted that communication cost for the distributed crossing minimization process is independent of the number of rows of the input matrix. On the other hand, most privacy preserving data mining algorithms incur a communication cost which is quadratic with the number of rows of the input matrix. Communication cost of the proposed framework scales linearly with the number of columns (attributes) and servers. This makes crossing minimization based biclustering algorithm a method of choice for scalable distributed implementations.

Broadcasting encrypted bicluster representatives would require $O(k_i \bar{\mathcal{A}}_{ci} \mathcal{N})$ message communications where each message is the size of RSA modulus. Clearly, overall communication cost scales linearly with the number of servers and attributes because the number of biclusters k_i at each server are assumed constant. This implies suitability of the proposed framework for large scale networks.

5   Experimental Framework

5.1   Experimental Setup

The proposed biclustering algorithm is implemented in C++. The cryptographic primitives are implemented using GNU MP Bignum Library. We used Message Passing Interface (MPI) for distributed memory implementation of the algorithm to mimic a distributed reliable network of servers. The algorithm's accuracy and performance was determined using synthetically generated data sets of large sizes. The experiments were performed on 10-Node Linux Cluster connected by Myrinet fast speed network. Each node in the cluster consists of a 0.8 GHz AMD Athlon processor with 512 MB RAM. MPICH-G2 1.2.4 is used to program our algorithm. The operating system is Gentoo Linux 2.6.9-r9 and compiler is the GNU gcc 3.3.6. In all the experiments the execution time was obtained through $MPI\_Wtime()$ and is reported in seconds. The execution time takes into account the time spent in reading data from Files in the process of initializing the structure.

5.2   Accuracy Evaluation

Our first set of experiments is aimed at comparing the accuracy of the non-secure crossing minimization based biclustering algorithm against non-secure K-means clustering algorithm. Both of these algorithms were run on a single processor

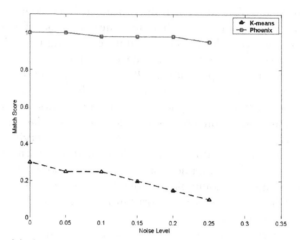

(a) Accuracy Comparison between Phoenix and K-means.

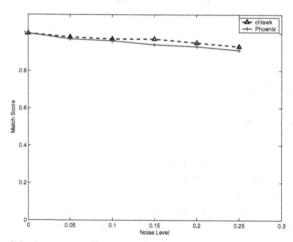

(b) Accuracy Comparison between Phoenix and cHawk.

Fig. 2. Accuracy Evaluation for Phoenix

with centralized data. To evaluate the accuracies of these algorithms, we use the measure (match score) similar to the score proposed by Prelic et al. [28] and Liu et al [29].

Let $M1, M2$ be two sets of bi-clusters. The match score of $M1$ with respect to $M2$ is given by

$$S(M_1, M_2) = \frac{1}{|M_1|} \sum_{A(I_1, J_1) \in M_1} max_{A(I_2, J_2) \in M_2} \frac{|I_1 \cap I_2||J_1 \cap J_2|}{|I_1 \cup I_2||J_1 \cup J_2|}$$

Let M_opt denote the set of implanted bi-clusters and M the set of the output bi-clusters of a (bi)clustering algorithm. $S(M_opt, M)$ represents how well each of the true bi-clusters is discovered by the algorithm.

We follow the approach used by Liu et al. [29] for synthetic data generation. To cater for the missing values in real life data, we add noise by replacing some elements in the matrix with random values. There are three variables b, c and γ in the generation of the bi-clusters. b and c are used to control the size of the implanted bi-cluster. γ is the noise level of the bi-cluster. The matrix with implanted constant bi-clusters is generated with four steps: (1) generate a 100×100 matrix A such that all elements of A are 0s, (2) generate ten 10×10 bi-clusters such that all elements of the bicluster are 1s, (3) implant the ten bi-clusters into A, (4) replace $\gamma(100 \times 100)$ elements of the matrix with random noise values (0 or 1).

Table 1. Parameter Settings for Phoenix and K-means

Method	Parameter Settings
Phoenix	$\delta = 0.5, Iterations = 5$, Function=$KL$-Divergence
K-means	$K = 10$

In the experiment, the noise level ranges from 0 to 0.25. The parameter settings used for the two methods are listed in Table 1. The results are shown in Figure 2(a). In the absence of noise, Phoenix can always find the implanted bi-clusters correctly. K-means algorithm significantly under performs as it misses many biclusters because of its inability to find local patterns. When the noise level is high, accuracy of Phoenix remains consistent throughout. Accuracy of K-means algorithm deteriorates with increase in noise. This lack of accuracy is due to its inability to disregard similarity between different objects based on noise.

Accuracy Loss due to Distributed Privacy Preserving Biclustering. Our second set of experiments are aimed at determining the effects of distributed privacy preserving primitives on the accuracy of the biclustering process. The resulting graph is shown in Figure 2(b). The graph illustrates the accuracy comparison between centralized crossing minimization based biclustering algorithm without security primitives (cHawk) [3] with Distributed Privacy Preserving Biclustering using Crossing Minimization (Phoenix). The graph shows that at zero noise, Phoenix is almost as accurate as cHawk. With increase in noise the accuracy degrades a little bit which is because of non-exact nature of bicluster comparison and merge operations of Phoenix. Note however that the accuracy of Phoenix remain considerably high even though it is a distributed implementation which uses a simple biclustering merge framework.

Performance Evaluation of Proposed Algorithm. Our second set of experiments were aimed at analyzing the performance in terms of execution times for Phoenix. For this purpose the algorithm was run on synthetic data sets with sizes ranging from 1000 (20×50) elements to 1000000 ($20,000 \times 50$) elements.

The number of columns in each case were fixed to be 50. The number of servers were fixed at four. We noted that increasing the number of servers did'nt have much impact on the execution times as it remain bounded by the size of the data.[1]

Figure 3 shows the performance in terms of execution time for varying set of problem sizes. The times reported are in seconds and indicate the time in performing the biclustering using crossing minimization. The curve in the graph represents the increase in execution time by increasing the problem size. As can be seen in the Figure 3, the time scales linearly with the increasing data sizes. Note that the curve seems non-linear, reason being the non-uniform scaling on the two axes. To clarify, we have attached the data table with the figure. For example the time taken to process 1,000 data elements is 2.8 seconds while the same for 50,000 data elements case is around 113 seconds. That implies that a fifty times increase in the data size result in an almost similar increase in the the biclustering time. Similar pattern can be observed on other data ranges. These experimental results verify our theoretical complexity analysis.

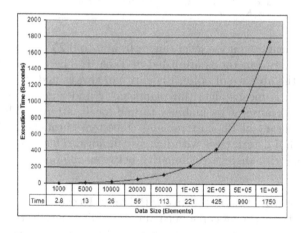

Fig. 3. Performance of Proposed Algorithm with increasing data size

6 Conclusions and Future Work

We have proposed a privacy preserving biclustering algorithm which is very efficient with respect to both computation and communication costs. This algorithm is suitable for those applications which require determination of local(subspace) patterns in horizontally partitioned data in privacy preserving manner. The proposed approach provides security among colluding malicious adversaries while also eliminating the requirement of trusted dealer. The algorithm is shown to

---

[1] This is because of the very fast network that we used for our experiments. Experimental evaluation over a realistic distributed network such as PlanetLab is a topic of our future research.

easily outperform a traditional K-means based clustering algorithm in finding fine grained patterns from the data. It also retains its accuracy in distributed setting among secure protocols for merging biclusters.

We are aiming at evaluating the proposed framework on a more practical network such as PlanetLab. Moreover, secure versions of distance metrics other than Euclidean distance are also being investigated. Because of the communication efficiency of the proposed framework, we are also evaluating its use for applications such as privacy preserving collaborative filtering in ubiquitous computing environments.

References

1. Clifton, C., Kantarcioglu, M., Doan, A., Schadow, G., Vaidya, J., Elmagarmid, A.K., Suciu, D.: Privacy-preserving data integration and sharing. In: DMKD, pp. 19–26 (2004)
2. Vaidya, J., Clifton, C.: Privacy-preserving k-means clustering over vertically partitioned data. In: The Ninth ACM SIGKDD International Conference on Knowledge Discovery and Data Mining, Washington, D.C., pp. 206–215 (2003)
3. Ahmad, W., Khokhar, A.: chawk: A highly efficient biclustering algorithm using bigraph crossing minimization. In: Second International Workshop on Data Mining and Bioinformatics, VDMB 2007, Vienna, Austria (In conjunction with VLDB 2007) (2007)
4. Sugiyama, K., Tagawa, S., Toda, M.: Methods for visual understanding of hierarchical system structures. IEEE Transaction on Systems, Man, and Cybernetics 11(2), 109–125 (1981)
5. Ahmad, W., Khokhar, A.: Secure aggregation in large scale overlay networks. In: IEEE Global Communications Conference (GLOBECOM), San Francisco, CA (2006)
6. Lin, X., Clifton, C., Zhu, M.: Privacy-preserving clustering with distributed em mixture modeling. Knowl. Inf. Syst. 8, 68–81 (2005)
7. Jha, S., Kruger, L., McDaniel, P.: Privacy preserving clustering. In: di Vimercati, S.d.C., Syverson, P.F., Gollmann, D. (eds.) ESORICS 2005. LNCS, vol. 3679, pp. 397–417. Springer, Heidelberg (2005)
8. Jagannathan, G., Wright, R.N.: Privacy-preserving distributed k-means clustering over arbitrarily partitioned data. In: KDD 2005: Proceeding of the eleventh ACM SIGKDD international conference on Knowledge discovery in data mining, pp. 593–599. ACM Press, New York (2005)
9. Cheng, Y., Church, G.: Biclustering of expression data. In: Proceedings of Intelligent Systems for Molecular Biology (2000)
10. Adam, N.R., Worthmann, J.C.: Security-control methods for statistical databases: a comparative study. ACM Comput. Surv. 21, 515–556 (1989)
11. Schlörer, J.: Security of statistical databases: multidimensional transformation. ACM Trans. Database Syst. 6, 95–112 (1981)
12. Agrawal, R., Srikant, R.: Privacy-preserving data mining. In: Proc. of the ACM SIGMOD Conference on Management of Data, pp. 439–450. ACM Press, New York (2000)
13. Lindell, Y., Pinkas, B.: Privacy Preserving Data Mining. In: Bellare, M. (ed.) CRYPTO 2000. LNCS, vol. 1880, Springer, Heidelberg (2000)

14. Vaidya, J., Clifton, C.: Privacy preserving association rule mining in vertically partitioned data (2002)
15. Kantarcioglu, M., Clifton, C.: Privacy-preserving distributed mining of association rules on horizontally partitioned data. IEEE Trans. Knowl. Data Eng. 16, 1026–1037 (2004)
16. Merugu, S., Ghosh, J.: Privacy-preserving distributed clustering using generative models. In: Proceedings of the Third IEEE International Conference on Data Mining ICDM 2003, Melbourne, FL (2003)
17. Kargupta, H., Datta, S., Wang, Q., Sivakumar, K.: On the privacy preserving properties of random data perturbation techniques. In: ICDM, pp. 99–106 (2003)
18. Dhillon, I.S.: Co-clustering documents and words using bipartite spectral graph partitioning. In: Proceedings of the Seventh ACM SIGKDD International Conference on Knowledge Discovery and Data Mining (KDD) (2001)
19. Ahmad, W., Khokhar, A.: An architecture for privacy preserving collaborative filtering on web portals. In: Proceedings of the Third International Symposium on Information Assurance and Security (2007)
20. George, T., Merugu, S.: A scalable collaborative filtering framework based on co-clustering. In: ICDM, pp. 625–628. IEEE Computer Society, Los Alamitos (2005)
21. Brglez, F., Stallmann, M.F., Ghosh, D.: Heuristics and experimental design for bigraph crossing number minimization. In: Goodrich, M.T., McGeoch, C.C. (eds.) ALENEX 1999. LNCS, vol. 1619, pp. 74–93. Springer, Heidelberg (1999)
22. Lafferty, J., Pietra, S., Pietra, V.: Statistical learning algorithms based on bregman distances. In: Proceedings of the Canadian Workshop on Information Theory (1997)
23. Ravikumar, P., Cohen, W.W., Fienberg, S.E.: A secure protocol for computing string distance metrics. In: Perner, P. (ed.) ICDM 2004. LNCS (LNAI), vol. 3275, Springer, Heidelberg (2004)
24. Malkin, M., Wu, T.D., Boneh, D.: Experimenting with shared generation of rsa keys. In: NDSS (1999)
25. Damgård, I., Jurik, M.: A generalisation, a simplification and some applications of paillier's probabilistic public-key system. In: Public Key Cryptography, pp. 119–136 (2001)
26. Afek, Y., Bremler, A.: Self-stabilizing unidirectional network algorithms by power supply. Chicago Journal of Theoretical Computer Science (1998)
27. Wagner, D.: Resilient aggregation in sensor networks (2004)
28. Prelic, A., Bleuler, S., Zimmermann, P., Wille, A., Buhlmann, P., Gruissem, W., Hennig, L., Thiele, L., Zitzler, E.: A systematic comparison and evaluation of biclustering methods for gene expression data. Bioinformatics (2006)
29. Liu, X., Wang, L.: Computing the maximum similarity bi-clusters of gene expression data. Bioinformatics 23, 50–56 (2007)

Allowing Privacy Protection Algorithms to Jump Out of Local Optimums: An Ordered Greed Framework

Rhonda Chaytor*

Dept. of Computer Science
Memorial University
St. John's, NL, Canada
rchaytor@cs.mun.ca

Abstract. As more and more person-specific data like health information becomes available, increasing attention is paid to confidentiality
and privacy protection. One proposed model of privacy protection is k-
Anonymity, where a dataset is k-anonymous if each record is identical to
at least (k-1) others in the dataset. Our goal is to minimize information
loss while transforming a collection of records to satisfy the k-Anonymity
model. The downside to current greedy anonymization algorithms is their
potential to get stuck at poor local optimums. In this paper, we propose
an *Ordered Greed Framework* for k-Anonymity. Using our framework,
designers can avoid the poor-local-optimum problem by adding stochastic elements to their greedy algorithms. Our preliminary experimental
results indicate improvements in both runtime and solution quality. We
also discover a surprising result concerning at least two widely-accepted
greedy optimization algorithms in the literature. More specifically, for
anonymization algorithms that process datasets in column-wise order,
we show that a random column ordering can lead to significantly higher
quality solutions than orderings determined by known greedy heuristics.

1 Introduction

One of the most important promises a physician makes to a patient is that of
confidentiality. At the same time, there is a need to release patient information
for research and surveillance. Today in our growing digital society, guaranteeing
patient privacy while providing researchers with worthwhile data has become
increasingly difficult. For example, suppose it is desirable to make a public release
of the patient dataset in Figure 1 for clinical research or health surveillance. In
the past, it was believed that *de-identification, i.e.,* removing obvious identifiers
like social security number and name, would be sufficient in the protection of
patient privacy. Recent studies, however, indicate that it is possible to re-identify
individuals, even if the data set is de-identified. A popular example can be found
in Sweeney [24], where the governor of Massachusetts was re-identified using

---

* This work has been supported by an NSERC Post Graduate Scholarship (Doctoral).

F. Bonchi et al. (Eds.): PinKDD 2007, LNCS 4890, pp. 33–55, 2008.

publicly available medical records and a voter registry (six people shared his birth date, three were male, and he was the only one in his 5-digit ZIP code).

One way to limit the risk of re-identification is to ensure that no information is distinctive to a particular individual, thereby making everyone anonymous. *k-Anonymity* [22,24] is a technique in which released records are made less specific, yet remain truthful. Unfortunately, we know that the problem of guaranteeing *k*-Anonymity is *NP*-hard [1,19].

The solution to this privacy protection problem is currently in high-demand, and since it is impossible for a software solution to produce an optimal solution efficiently, researchers are concentrating on faster heuristic approaches that produce good sub-optimal results. Although this is a new research area, it appears that genetic algorithms have already been dismissed as viable solutions due to reports of lower accuracy and much higher runtime [4,7,26]. In our opinion, these two issues can be addressed by appropriate techniques from genetic algorithms research. In this paper, we propose a new framework that allows algorithm designers to avoid the poor-local-optimum problem by adding stochastic elements to their greedy algorithms. Unlike previous bit string genetic algorithms, this framework is based on a permutation problem representation and a genetic algorithm approach called *ordered greed*. We compare the runtime and solution quality of our new framework to that of previous genetic algorithms and discover a surprising result concerning at least two widely-accepted greedy optimization algorithms in the literature. More specifically, given that anonymization algorithms often process datasets in column-wise order, we show that a random column ordering can significantly lead to higher quality solutions than orderings determined by known greedy heuristics. We discuss our implementation of one of these greedy optimization algorithms, `Datafly` [23], and compare its performance as a stand-alone algorithm to its performance as a module inside our new framework. As an added bonus, we complete nearly all the future work listed by leading experimental researchers in this area [26]:

- avoiding getting stuck at poor local optimums
- handling data suppression
- local recoding (not necessary to generalize all identical data values together)
- generalizing numeric attributes without hierarchies

This paper is organized as follows. Section 2 provides overviews and describes the previous work of the main themes of this paper: *k-Anonymity* and *Genetic Algorithms*. Section 3 describes our new framework in detail, including an overview, usage requirements, and how a greedy optimization algorithm, such as `Datafly` [23], can be configured to work as a module inside our framework. Section 4 describes specific implementation details for the Ordered Greed Framework, `Datafly`, and a traditional genetic algorithm modeled after existing genetic algorithms for *k*-Anonymity [8,17]. Section 4 also shows our quality and runtime results. Section 5 provides insight into our results and Section 6 gives a brief summary and suggests several directions for future research.

2   Background

This section reviews the two major themes of this paper: *k-Anonymity* and *Genetic Algorithms*. For both of these themes, we provide an overview describing fundamental concepts and a discussion of previous work.

2.1   *k*-Anonymity

Recall from Section 1 that *k*-Anonymity is the privacy-protection property that prohibits re-identification by making person-specific records less specific. Content found in this section is from [6, Section 2.2] and as the purpose of this section is to provide necessary background for understanding the framework in the next section, what follows is not a complete treatment of *k*-Anonymity.

Given an $n \times m$ dataset D, such as the one depicted in Figure 1, let rows represent people and columns represent a set $A = \{a_1, a_2, \ldots, a_m\}$ of human attributes.[1] A user-defined *quasi-identifier*, $Q = \{q_1, q_2, \ldots, q_h\}$, $h \leq m$, specifies which columns contain *personal information*, *i.e.*, information that is common enough to exist in other datasets, where linking can occur. The contents of the other $(m - h)$ columns make up *private information*, *i.e.*, information that is unique to this dataset, where linking cannot occur. For example, in Figure 1 the contents of `Problem` are considered private information and the contents of all other columns are considered personal information.

	Age	Work Class	Education	Marital Status	Occupation	Race	Sex	Native Country	Problem
1	39	State-gov	Bachelors	Never-married	Adm-clerical	White	Male	United-States	obesity
2	50	Self-emp-not-inc	Bachelors	Married-civ-spouse	Exec-managerial	White	Male	United-States	chest pain
3	38	Private	HS-grad	Divorced	Handlers-cleaners	White	Male	United-States	flu
4	53	Private	11th	Married-civ-spouse	Handlers-cleaners	Black	Male	United-States	cancer
5	28	Private	Bachelors	Married-civ-spouse	Prof-specialty	Black	Female	Cuba	obesity
6	37	Private	Masters	Married-civ-spouse	Exec-managerial	White	Female	United-States	obesity
7	49	Private	9th	Married-spouse-abse	Other-service	Black	Female	Jamaica	flu
8	52	Self-emp-not-inc	HS-grad	Married-civ-spouse	Exec-managerial	White	Male	United-States	chest pain
9	31	Private	Masters	Never-married	Prof-specialty	White	Female	United-States	cancer
10	42	Private	Bachelors	Married-civ-spouse	Exec-managerial	White	Male	United-States	obesity

Fig. 1. Patient Dataset

---

[1] This is a partial dataset consisting of the first ten records of the Adult Database from the UCI Machine Learning Repository (`http://mlearn.ics.uci.edu/MLSummary.html`); however, for simplicity we assume that it is complete and use it as a running example throughout this paper.

To achieve k-Anonymity, the entry values of the quasi-identifier are general-ized (*i.e.*, made less specific) so that groups of at least k people look identical in terms of their personal information. Their private information, like `cancer` from Figure 1, remains unmodified. Generalization can be accomplished using *domain generalization hierarchies* (*DGH*'s). To illustrate how they are used, consider the Work Class *DGH* from Figure 2 (adapted from [7]). At the bottom of the Work Class *DGH* no generalization is applied; however, as we traverse up the hierarchy levels, we notice that the work class becomes increasingly more general. The generalization at the top of the *DGH*, where no information is re-vealed, is called suppression. In this paper, we represent a suppressed entry using a star symbol ($\star$). The problem definition (adapted from SOL-k-ANONYMITY ON ENTRIES [6]) can now be stated as:

k-ANONYMITY

Instance: An $n \times m$ dataset D, a quasi-identifier Q, DGH's for each
                  column in Q, and a positive integer k.
Solution: The least generalized k-anonymous dataset $g(D)$.

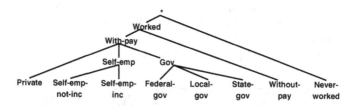

Fig. 2. Work Class DGH

We assume all *DGH*'s to be *single-level DGH*'s, where each domain value maps to $\star$ (see 4-level and single-level *DGH*'s in Figure 4). The reason for using single-level *DGH*'s is discussed further in Section 4. Given these single-level *DGH*'s, the mapping of the patient dataset in Figure 1 to a 2-anonymous patient dataset is shown in Figure 3.

2.2   Previous k-Anonymity Work

Three main k-Anonymity experimental research areas have been quickly advanc-ing in the past few years:

- **Enhancement:** ensuring no other inference attacks are possible after apply-ing k-Anonymity, e.g., (α, k)-Anonymity [28], distributed k-Anonymity [9], t-Closeness [15], p-Sensitive k-Anonymity [25], ℓ-Diversity [18].
- **Utility:** making the anonymized dataset more useful for the goal at hand, e.g., classification [8], personalized privacy preservation [29], extra aggregate information in the form of marginals [10], pattern discovery [3], target work-loads of selection queries and data mining tasks [14], important columns [30], ad hoc aggregate analysis [11].

	Age	Work Class	Education	Marital Status	Occupation	Race	Sex	Native Country	Problem
1	*	*	*	*	*	White	Male	United-States	obesity
3	*	*	*	*	*	White	Male	United-States	flu
2	*	*	*	Married-civ-spouse	Exec-managerial	White	Male	United-States	chest pain
8	*	*	*	Married-civ-spouse	Exec-managerial	White	Male	United-States	chest pain
10	*	*	*	Married-civ-spouse	Exec-managerial	White	Male	United-States	obesity
4	*	Private	*	*	*	Black	*	*	cancer
5	*	Private	*	*	*	Black	*	*	obesity
7	*	Private	*	*	*	Black	*	*	flu
6	*	Private	Masters	*	*	White	Female	United-States	obesity
9	*	Private	Masters	*	*	White	Female	United-States	cancer

Fig. 3. 2-Anonymous Patient Dataset

- **Algorithmics:** proposing new heuristics and data structures, *e.g.,* Datafly [23], k-Minimal Generalization [22], Bottom-Up Generalization [26], k-Optimize [4], Top-Down Specialization [7], Incognito [12], Top-Down Greedy Strict Multidimensional Partitioning [13], Greedy k-Member Clustering [5], Genetic Algorithms [8,17].

While continued research in each of the above areas is necessary for developing practical solutions for privacy protection, this paper concentrates on limitations of current Algorithmics work. In particular, genetic algorithms are limited by their long runtimes (*e.g.,* 18 hours [8]) and greedy optimization algorithms are limited by their potential of getting stuck at local optimums (*e.g.,* [26]). We propose to overcome these limitations by benefiting from the best of both worlds; we use genetic algorithm elements to search for good solutions and greedy optimization to evaluate the quality of these solutions. Appropriate techniques from genetic algorithms research are discussed next and our framework is described in Section 3.

2.3   Genetic Algorithms

In this section, similar to our treatment of k-Anonymity, we only provide the necessary background for understanding the genetic algorithms and Ordered Greed Framework discussed in this paper. The following genetic algorithm elements are taken from [20, Chapter 1]:

- **Population:** individual members of a *population* are typically bit strings and are called *chromosomes*. At the beginning of the genetic algorithm, the population is randomly initialized with a fixed number (*i.e., population size*) of chromosomes. A chromosome is essentially a point in the search space of candidate solutions.

- **Fitness:** a genetic algorithm evaluates the *fitness* of a chromosome, depending on how well it solves the given problem using a *fitness function*. After all fitnesses are evaluated for a population, usually the average, best, and worst fitnesses are recorded. A *fitness landscape* represents the space of all possible chromosomes and their associated fitnesses.
- **Selection:** this operator selects *parent* chromosomes from the population for *reproduction*. It is likely that chromosomes with the best fitnesses are selected.
- **Crossover:** after randomly choosing one or more positions, this operator combines selected parents to create *offspring* chromosomes. Based on offspring fitness and the *replacement* method used, offspring may (1) be added to the population, (2) replace existing members of the population, or (3) not be added to the population at all. Crossover can be seen as a way of creating *diversity*, by moving the population around on the fitness landscape to explore new candidate solutions and possibly jump out of any local optimums.
- **Mutation:** after a chromosome is chosen with a certain probability, this operator chooses one or more positions and mutates (*e.g.*, changing a 0 to a 1 in a bit string) the chosen chromosome and replaces it in the population. The probability is fixed in advance and is called the *mutation rate*. Mutation is another way of introducing diversity into the population.

A good genetic algorithm finds a balance between exploring the fitness landscape and exploiting good chromosomes. In general, a genetic algorithm works as follows:

1. Randomly create a population of chromosomes
2. For each chromosome, evaluate fitness
3. Select parents
4. Crossover parents to produce offspring
5. For offspring, evaluate fitness
6. Place offspring into population if necessary
7. With a mutation rate, mutate chromosomes
8. Repeat steps 2-7 until termination criteria is satisfied

Each iteration of steps 2-7 is called a *generation* and the entire set of generations is called a *run*. Given the stochastic nature of genetic algorithms, researchers usually report statistics (*e.g.*, average, best, and worst fitness) averaged over a number of different runs on the same problem.

2.4 Previous Genetic Algorithms Work

Iyengar [8] uses a binary bit string (($n - 1$) bits for each column, where n is the number of values in that column's domain) to represent a generalization scheme, where a 1 indicates an endpoint of an interval where a generalization occurs. For example, the top of Figure 4 shows a bit string representing this sort of generalization scheme.

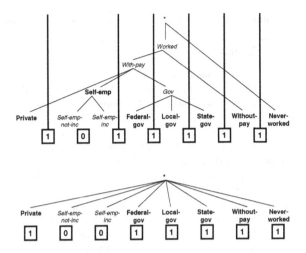

Fig. 4. Different Ways to Map a Bit String to a Generalization Scheme

After applying a generalization scheme, if a row still does not look like $(k-1)$ other rows, the entire row is suppressed and a penalty equal to the number of suppressions in that row is added to the fitness function. The fitness function used is the generalization loss metric (LM):

$$LM = \left(\sum_a \frac{1}{N_a - 1} \sum_r P_{a,r} - 1 \right) + penalties$$

where $P_{a,r}$ is the number of leaf nodes in the subtree of column a's DGH rooted at the generalized value for the value in row r. For example, assume column a is Work Class and the value in row r is Self-emp-not-inc. Referring again to the top of Figure 4, since the 0 bit implies that Self-emp-not-inc is generalized to Self-emp and there are 2 values in the subtree rooted at Self-emp, $P_{a,r} = 2$. In this example $LM = (2-1)/(8-1) = 1/7$.

Under Iyengar's scheme, there are only 9 legal bit strings for the Work Class DGH shown in the top of Figure 4. Consider bit string 1 0 1 0 1 1 1, which is illegal because if Federal-gov and Local-gov are generalized to Gov, then so should State-gov. To avoid these illegal bit strings, a repair method is used, which maps illegal bit strings to legal bit strings.

To test their algorithm, they used 30162 of the 32561 rows (due to missing-value limitations in their system) from the UCI Machine Learning Repository Adult Database benchmark and the 8 columns described in Table 1. Each run used a population size of 5000 and terminated after 0.5 million iterations. The mutation rate was set at 0.002 and it took 18 hours to run.

To speed up Iyengar's runtime, Lunacek, Whitley, and Ray [17] designed a crossover operator that avoided illegal offspring, used a smaller population size, changed the termination condition, and avoided mutation altogether. Although they do not actually report their runtime, they do show graphically that their

Table 1. Mapping Columns from the Adult Database to a Permutation

Map Col		Domain	Size
0	Age	Continuous from 17-90	74
1	Work Class	Private, Self-emp-not-inc, Self-emp-inc, Federal-gov, Local-gov, State-gov, Without-pay, Never-worked	8
2	Education	Preschool, 1st-4th, 5th-6th, 7th-8th, 9th, 10th, 11th, 12th, HS-grad, Some-college, Assoc-acdm, Assoc-voc, Bachelors, Prof-school, Masters, Doctorate	16
3	Marital Status	Married-civ-spouse, Married-AF-spouse, Divorced, Separated, Widowed, Married-spouse-abse, Never-married	7
4	Occupation	Exec-managerial, Prof-specialty, Sales, Adm-clerical, Tech-support, Craft-repair, Machine-op-inspct, Handlers-cleaners, Transport-moving, Priv-house-serv, Protective-serv, Armed Forces, Farming-fishing, Other-service	14
5	Race	White, Asian-Pac-Islander, Amer-Indian-Eskimo, Other, Black	5
6	Sex	Female, Male	2
7	Native Country	United-States, Outlying-US(Guam-US, Canada, Mexico, Honduras, Guatemala, Nicaragua, El-Salvador, Ecuador, Peru, Columbia, Puerto-Rico, Dominican-Republic, Jamaica, Cuba, Haiti, Trinadad& Tobago, France, England, Ireland, Scotland, Holand-Netherlands, Italy, Greece, Portugal, Yugoslavia, Hungary, Germany, Poland, Philippines, Thailand, Cambodia, Vietnam, Laos, India, Japan, China, Hong, Taiwan, South, Iran	41

genetic algorithm is more efficient and effective than Iyengar's. As a final point on previous work, Lunacek, Whitley, and Ray admit that Iyengar's representation (which they adopt) is problematic: operators produce illegal bit strings and representation introduces bias. In the next section, we describe our Ordered Greed Framework, which uses a non-traditional permutation representation to avoid these problems.

3 Ordered Greed Framework

In this section, we propose a framework that incorporates both the speed of greedy optimization algorithms and improvement potential of genetic algorithms. This framework allows anonymization algorithms to arrive at better local optimums than existing algorithms, which was listed as future work by leading k-Anonymity research [26]. After a brief overview and explanation of usage requirements, we demonstrate using Sweeney's Datafly algorithm [23] how algorithm designers may use our Ordered Greed Framework.

3.1   Framework Overview

Our framework is structured the same as the basic genetic algorithm outlined in Section 2.3. First random chromosomes are created and the fitness of each chromosome is computed using a greedy optimization algorithm (*e.g.*, Datafly). Based on their fitnesses, chromosomes are selected as parents for crossover and one or more new chromosomes are produced. The fitness of each new offspring chromosome is then computed using the same greedy optimization algorithm as before. A new population is created according to a particular replacement strategy and each member of this new population has a specified chance of being mutated. Fitnesses are again determined using the greedy optimization algorithm and the process continues until the termination criteria is satisfied. This framework is depicted in Figure 5.

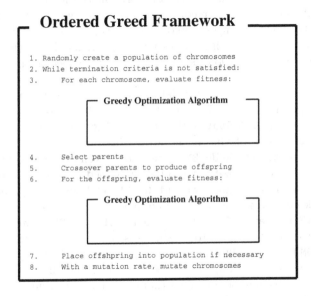

Ordered Greed Framework

```
1. Randomly create a population of chromosomes
2. While termination criteria is not satisfied:
3.      For each chromosome, evaluate fitness:

              Greedy Optimization Algorithm



4.      Select parents
5.      Crossover parents to produce offspring
6.      For the offspring, evaluate fitness:

              Greedy Optimization Algorithm



7.      Place offshpring into population if necessary
8.      With a mutation rate, mutate chromosomes
```

Fig. 5. Ordered Greed Framework

3.2   Framework Usage Requirements

An *ordered greed* genetic algorithm [2] processes a population of *permutations* and fitness is greedily evaluated based on the order in which parts of a problem are solved. For example, a permutation in graph-coloring may specify the order in which vertices are colored. After vertex v is colored, any vertex which follows v in the ordering and is adjacent to v must not be colored the same as v.

To use our Ordered Greed Framework, the algorithm that evaluates fitness (see lines 3 and 6 in Figure 5) must be

1. a greedy optimization algorithm, and
2. dependent on column ordering.

Each chromosome is a permutation, where permutation elements represent column labels and the permutation itself represents a column ordering. This is a very natural representation for k-ANONYMITY and there are algorithms in the literature which depend on column order when exploring the search space. For example, column orders can be based on the most distinct values [23] or the widest range of values [13]. We demonstrate in the next section how algorithms like these may use our framework to introduce randomness and jump out of local optimums.

3.3   Framework in Action: Adding Randomness to Datafly

Sweeney's Datafly algorithm [23] satisfies the framework usage requirements described in the previous section; it greedily generalizes a column based on the largest domain size. Essentially, Datafly works in the following way: given a permutation of column labels, greedily group records together that have identical values for the columns encountered so far in the ordering. Maintain groups of at least size k by suppressing values where necessary. For each consecutive column label in the permutation, possible row groupings become more and more restrictive.

For example, consider the Patient dataset in Figure 1 and assume the columns are mapped to permutation 5 6 7 3 4 0 1 2, $i.e.$, the column ordering is Race, Sex, Native Country, Marital Status, Occupation, Age, Work Class, Education.[2] To evaluate the fitness of this example, we first execute the query SELECT COUNT(*) FROM Patient GROUP BY Race. We can tell from COUNT(*) that $k \geq 2$, so the number of values that have to be suppressed $= 0$. Next we execute the query SELECT COUNT(*) FROM Patient GROUP BY Race, Sex. The patients that are Black and Male or Female are no longer in groups of size $k \geq 2$, so we execute UPDATE Patient SET Male=* AND Female=* WHERE Race=Black and the number of values that have to be suppressed $= 3$. Now we execute SELECT COUNT(*) FROM Patient GROUP BY Race, Sex, Native Country and continue in the same way until we group by all the columns in permutation order. Given that the total possible number of values that could have been suppressed is $n \times m$ (the number of rows multiplied by the number of columns), the fitness for permutation 5 6 7 3 4 0 1 2 $= 10 \times 8 - (0+3+3+7+7+10+5+8) = 80-43 = 37$, which is the number of entries in the resulting k-anonymized version of D that are not suppressed, as shown in Figure 3.

Previous heuristic approaches have made use of optimizations like pruning [4, 26], pre-computation [12], and crossovers that preserve validity [17]. Since UPDATE queries are very expensive, we avoid them using a $tree$ optimization. The tree is an internal data structure created from a single GROUP BY query, where each node in the tree has an ID, a $count$, zero or more $children$, and exactly one $parent$. For example, the tree corresponding to the dataset in Figure 1

---

[2] Note that this permutation was chosen arbitrarily and was not determined by the Datafly heuristic ($i.e.$, column-wise ordering based on domain size in descending order) for the Adult dataset. According to the domain sizes from Table 1, the permutation determined by Datafly for the Adult dataset is 0 7 2 4 1 3 5 6.

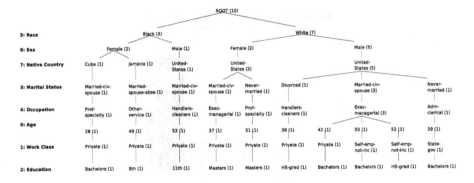

Fig. 6. Internal Data Structure for Patient Dataset given permutation 5 6 7 3 4 0 1 2

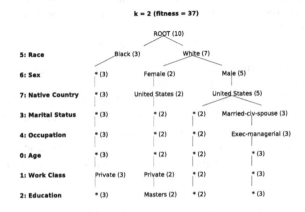

Fig. 7. 2-Anonymous Internal Data Structure for Patient Dataset given permutation 5 6 7 3 4 0 1 2

and permutation 5 6 7 3 4 0 1 2 is created after executing the query SELECT *, COUNT(*) FROM Patient GROUP BY Race, Sex, Native Country, Marital Status, Occupation, Age, Work Class, Education. This tree is shown in Figure 6 and the 2-anonymous version is shown in Figure 7.

In addition to the above tree optimization for Datafly, we can optimize the Ordered Greed Framework by storing a *lookup table* that contains the fitnesses evaluated in each generation; instead of blindly re-calculating fitnesses in steps 3 and 6 of Figure 5, we first check to see if it has already been evaluated and stored in the lookup table. These optimizations (*i.e.*, tree and lookup) significantly improve the overall runtime (see Figure 8).

The optimized Datafly algorithm is given in Figures 9 and 10. Lines 1–5 of Figure 9 determine column domain sizes and place the columns in descending domain-size order. Line 6 performs a GROUP BY query on all columns in this order. Lines 7–19 build a tree, such as the one depicted in Figure 6, from the result of this query. After enforcing anonymity in line 20 (described next), lines

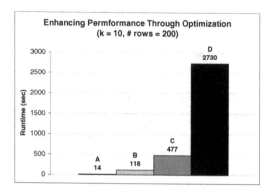

Fig. 8. How Optimization Improves Performance. A is the runtime when both optimizations are used, B is the runtime when only the tree optimization is used, C is the runtime when only the lookup optimization is used, and D is the runtime when neither of the optimizations are used.

21–23 count all the suppressions in the tree, calculate fitness (*i.e.*, the number of values not suppressed), and return this fitness.

The pseudocode for enforcing anonymity in Figure 10(a) describes a recursive function with the majority of work occurring in line 4, where a helper function is called. This helper function is also a recursive function and is given in Figure 10(b). Line 1 of this helper function initializes three variables: (1) *totUnderk* sums the counts of nodes that have suppressed values, (2) *minOverk* keeps track of the smallest count of nodes with values other than ⋆, and (3) *minIndex* records the index of the node with the count of size *minOverk*. In lines 2–10, for each child of an input node, if it is necessary to suppress a value, then appropriate updates are made to *totUnderk*, *minOverk*, and *minIndex*. In lines 11–14, if there were suppressions, but still not enough to satisfy k-Anonymity, then the value of the node indexed at *minIndex* is suppressed. Finally, lines 15–16 merge nodes and update counts for nodes with matching values.

To use this algorithm in our framework, we simply omit lines 1–4 in Figure 9, which determine the domain size of each column and the column ordering based on domain sizes. As the framework randomly generates column orderings instead of generating a particular ordering based on a greedy heuristic, these lines of code are no longer necessary. Notice that besides the framework's ability to jump out of local optimums, the `Datafly` algorithm involves suppressing data, local recoding[3], and generalizing/suppressing numeric attributes without hierarchies, which are extensions listed as future work by leading researchers (see Section 1).

---

[3] An example of local recoding (*i.e.*, it is not necessary to generalize all identical data values together [12]) is apparent from the `Work Class` nodes of the tree in Figure 7, since instances of `Private` exist, even though the same value is suppressed for `White Males` from the `United States`. Note that Sweeney's version of `Datafly` [23] is only capable of global recoding, (*i.e.*, if one instance of `Private` has to be suppressed, then all other instances of `Private` must be suppressed as well.)

4   Experimental Setup

We designed experiments to test the following hypotheses:

1. The Ordered Greed Framework has a faster runtime and can produce higher quality results (*i.e.*, larger number of entry values that are not suppressed)

```
1.  for each column i in D
2.  data = result of SELECT COUNT(*) FROM D GROUP BY i's name
3.  size[i] = COUNT(*)
4.  permutation order = column order determined by sorting size[i] in descending order
5.  sql = column names in permutation order
6.  data = result of SELECT sql, COUNT(*) FROM D GROUP BY sql
7.  root = new node (ID="ROOT", count=0, children=null, parent=null)
8.  for each record in data
9.  for each entry in this record for which a node does exist
10. do nothing
11. for each entry in this record for which a node does not yet exist
12. if this entry corresponds to a leaf node
13. c = COUNT(*)
14. else
15. c = 0
16. currentNode = new node (ID=this entry, count=c, children=null, parent=root)
17. root.addChild(currentNode)
18. root = currentNode
19. propagateCounts(root)
20. EnforceAnonymity(root, k)
21. setNumSuppressions(root)
22. fitness = numRecords * numCols - numSuppressions
23. return(fitness)
```

Fig. 9. Pseudocode for Datafly Algorithm

(a)

```
1.  node.getChildren()
2.  if node is a leaf node
3.  return
4.  EnforceAnonymityHelper(node, k)
5.  node.getChildren()
6.  for each child
7.  EnforceAnonymity(child, k)
```

(b)

```
1.  minIndex = 0, totUnderk = 0, minOverk = INFINITY
2.  node.getChildren()
3.  for each child
4.  if (child.getCount() < k)
5.  totUnderk = totUnderk + child.getCount()
6.  child.setID("*")
7.  else
8.  if (child.getCount() < minOverk)
9.  minOverk = child.getCount()
10. minIndex = this child's index
11. if ((totUnderk < k) && (totUnderk > 0))
12. minNode = child with minIndex
13. totUnderk = totUnderk + minNode.getCount()
14. minNode.setID("*")
15. if (totUnderk > 0)
16. MergeLikeChildren(node)
```

Fig. 10. Pseudocode for (a) EnforceAnonymity and (b) EnforceAnonymityHelper Algorithms

than previous genetic algorithms (*i.e.*, Iyengar's [8] and Lunacek, Whitley, and Ray's [17]).

2. The Ordered Greed Framework that evaluates fitness using an existing greedy optimization algorithm (*e.g.*, Sweeney's `Datafly` [23]) can produce higher quality results than the existing greedy optimization algorithm can alone. Obviously, `Datafly` is faster than `Datafly` + the Ordered Greed Framework.

In this section, we provide implementation details and results for these experiments. As the number of variables should be minimal when comparing one algorithm to another, we abide by the following three guidelines:

- **Single-level DGH's:** to reduce the dependencies on hierarchies and overhead in dealing with illegal chromosomes. For example, in our implementation of previous genetic algorithms, we change the bit string mapping from the top of Figure 4 to the mapping shown at the bottom of Figure 4. The key observation here is that a generalization scheme which changes a value to $\star$ has a loss of $LM = (8 - 1)/(8 - 1) = 1$ (*i.e.*, for single-level DGH's, LM = number of $\star$'s in the anonymized dataset). Deciding if one hierarchy is better than another is a utility issue, not an algorithmics issue.
- **Loss Metric:** to highlight algorithmic performance. If the algorithm works for one general metric at the cell-level, it can be extended to work for other metrics.
- **Missing Data:** to show limitations. Rather than delete records that have missing values (*e.g.*, in the Adult dataset, a missing value is represented by a question mark), we treat these values the same as any other value in the domain. As we expect real-world datasets to have missing data, processing such a dataset *as is* allows for less information loss and is more practical.

4.1   Implementation

Like previous work, we use 8 columns from the Adult Database benchmark (see Table 1) and assume it fits into memory. Our code is written in Java and we use a mySQL database. All experiments were run on an AMD Athlon 1809 MHz machine with a cache size of 512 KB and 895 MB of free memory.

Our algorithm implementations are named in the following way: Sweeney's `Datafly` algorithm is called `Datafly`, the *O*rdered *G*reed *F*ramework which evaluates fitness using `Datafly` is called `Datafly+OGF`, or just `OGF`, and the *trad*itional *G*enetic *A*lgorithmic approach used in previous work is called `tradGA`. Implementation details for each of these algorithms is discussed next.

- **Datafly:** Refer to pseudocode and description in Section 3.3.
- **OGF:** For a broad overview of `OGF`, see Section 3.1. All experiments use the implementation described in Table 2; the number of fitness evaluations, number of runs, and selection and replacement methods are the same as in Lunacek, Whitley, and Ray [17] and we performed a series of experiments to determine suitable values for population size and mutation rate, as well

as which operators to use. Note that the maximum number of fitness evaluations = (*population size+number of offspring*)× *number of generations* = $(10 + 1) \times 15000 = 165000$. Through experimentation, we found that such a large number of fitness evaluations was unnecessary, as the best solution did not change for a majority of the 15000 generations. We added the *no change in the best solution after 10 generations* condition to the termination criteria (to avoid useless evaluations) and as a result, reduced the number of generations from 15000 to 11 - 75 generations (*i.e.,* the maximum number of fitness evaluations = $(10 + 1) \times 75 = 825$). Setting the maximum number of consecutive generations without improvement is a widely used termination criteria in genetic algorithms research (*e.g.,* [16, 21]).

– **tradGA:** When we compare OGF to tradGA, we are essentially comparing permutation and bit string representations. Therefore, we use the same OGF implementation as above and only modify it enough to accommodate a bit string representation. In particular, we modify the internal data structure (see Figure 6) to always use permutation order 0 1 2 3 4 5 6 7. We also modify the EnforceAnonymityHelper pseudocode (see Figure 11) and the crossover and mutation operators (see Figure 12). The modified helper algorithm works in the following way. For each possible value of each domain, lines 1–6 of Figure 11 check if suppression is necessary. Then, in line 7, nodes with matching values are merged and their counts are updated.

Table 2. Ordered Greed Framework Implementation Details

Population size	10
Number of runs	15
Termination criteria	Terminate after 15000 generations or when no improvement in best fitness after 10 generations
Selection	*Rank*; select top two parents (highest fitness)
Crossover	*Ordered* for permutation
Replacement	Offspring replaces worst chromosome if better
Mutation	*Swap* for permutation
Mutation rate	10 %

```
1. for each value in each domain
2. if (corresponding bit==0)
3. node.getChildren()
4. for each child
5. if (child.getID() == this value)
6. child.setID("*")
7. MergeLikeChildren(node)
```

Fig. 11. Pseudocode for EnforceAnonymityHelper Algorithm for Bit Strings

4.2 Results

This section shows the various solution quality results for Datafly, OGF+Datafly, and tradGA, where quality is based on fitness (*i.e.,* the number of entry values

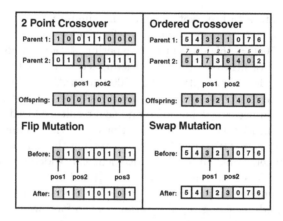

Fig. 12. Examples of the Crossover and Mutation Operators

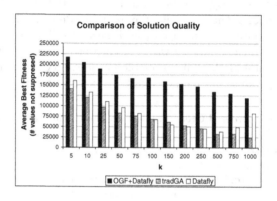

Fig. 13. Comparing Solution Quality for Each Value of k

not suppressed), as well as runtime results. We report average results over 15 runs for $k = 5, 10, 25, 50, 75, 100, 150, 200, 250, 500, 750, 1000$.

- **Solution Quality:** For each algorithm, Figure 13 shows a graph which plots the best fitness over all runs and all generations for increasing values of k.
- **Quality Improvements:** Table 3 shows how Datafly's solution quality improves using our Ordered Greed Framework. Table 4 lists the best and worst solutions found by OGF+Datafly and their corresponding fitnesses for each value of k.
- **Search Space Exploration:** Figure 14 displays a graph that plots, for OGF+Datafly and tradGA, the best, average, and worst fitnesses for an increasing number of generations when $k=5$.
- **Runtime:** Figure 15 shows a graph which plots, for each algorithm, the average runtime over all runs and all generations for increasing values of k.

Table 3. OGF Improves Solution Quality. This table lists the percent of dataset entry values not suppressed for Datafly and Datafly+OGF and the percent improvement for each value of k.

k	Datafly	Datafly+OGF	Improvement
5	61.68%	84.80%	23.12%
10	51.08%	80.76%	29.68%
25	42.29%	75.42%	33.13%
50	36.87%	71.44%	34.56%
75	31.43%	68.66%	37.22%
100	25.99%	67.49%	41.50 %
150	21.01%	65.34%	44.34%
200	19.29%	62.92%	43.62%
250	17.24%	60.91%	43.67%
500	14.72%	55.60%	40.87%
750	18.71%	53.53%	34.82%
1000	31.45%	49.63%	18.18%

Table 4. OGF: Summary of Best and Worst Solutions

k	Best Fitness	Best Solution	Worst Fitness	Worst Solution
5	220892	1 5 6 7 3 2 0 4	150191	2 0 4 3 1 5 7 6
		2 5 1 7 0 6 3 4		
		3 5 6 7 1 2 0 4		
		6 5 1 7 0 2 3 4		
		6 5 1 7 3 2 0 4		
10	210362	3 5 6 7 1 2 0 4	118333	4 0 2 6 1 3 5 7
25	196453	5 3 7 6 1 4 2 0	80917	0 4 7 5 2 1 3 6
50	186085	3 7 6 5 1 4 0 2	65436	0 4 2 7 5 1 3 6
75	178841	6 5 7 3 2 1 4 0	59086	0 4 7 2 3 5 6 1
100	175803	7 6 5 1 3 4 0 2	60163	0 4 3 2 5 6 1 7
		7 6 5 1 3 4 2 0		
150	170209	6 7 5 3 1 4 0 2	54718	0 2 1 3 5 6 4 7
		0 7 5 3 1 4 6 2		7 6 2 4 1 0 3 5
200	163888	5 6 7 1 3 4 0 2	50132	0 4 2 3 5 7 6 1
250	158651	6 5 7 3 1 2 0 4	45340	0 2 7 5 6 4 3 1
		6 5 4 1 3 2 0 7		
		6 3 4 1 5 2 0 7		
500	144819	5 6 7 3 0 1 2 4	31252	0 4 1 3 5 2 7 6
		5 6 7 3 2 1 0 4		0 4 1 2 6 3 5 7
750	139442	6 7 5 1 3 4 2 0	35263	0 4 5 6 1 2 3 7
		5 7 6 1 3 4 2 0		
1000	129271	7 5 0 6 3 1 2 4	44439	4 3 2 6 1 5 7 0
		7 5 4 6 3 1 2 0		4 3 0 2 5 1 6 7
		7 5 1 6 3 4 2 0		4 5 3 7 6 1 2 0

5   Analysis and Discussion

This section discusses our experimental results in detail. In Section 5.1, we compare the quality of `Datafly+OGF`, `tradGA`, and `Datafly`. In Section 5.2, we discuss `Datafly` quality improvements using `OGF`. In Section 5.3, we analyze population data to understand why `Datafly+OGF` obtains higher quality solutions than `tradGA`, and in Section 5.4, we compare the runtime of each algorithm. For all results, significance is based on standard paired t-tests with probability $= 0.01$.

(a)

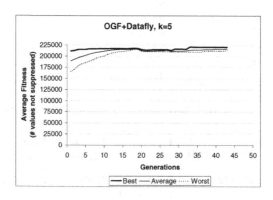

(b)

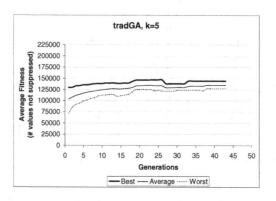

Fig. 14. `OGF` vs. `tradGA`: Average Fitness Over 15 Runs for Each Generation

5.1   `Datafly+OGF` Produces the Highest Quality Solutions

Recall that quality is based on fitness (*i.e.*, the number of entry values that are not suppressed). Notice that in Figure 13, (`Datafly+OGF`) produces statistically higher quality results than those of `Datafly` and `tradGA`. There is no significant difference in quality for `Datafly` and `tradGA` over all values of k; however, it

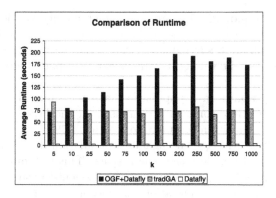

Fig. 15. Comparing Runtime for Each Value of k

is interesting that `Datafly` outperforms `tradGA` for very low and very high k values. We speculate this phenomenon may be related to the distribution of data. While further investigation is required, support for this speculation comes from the fact that `Datafly` at $k = 1000$ outperforms $k = 750$ by almost a factor of 2. Given that `Datafly` is a deterministic greedy algorithm, genetic algorithm elements like number of fitness evaluations and number of runs do not play a role in the above phenomenon. As we will see next, a column ordering, such as 0 7 2 4 1 3 5 6 (determined by `Datafly` for the Adult dataset), may produce good results for some values of k and may not for other values of k.

5.2   `OGF` Allows `Datafly` to Jump Out of Local Optimums

Table 3 shows that a greedy optimization algorithm for k-Anonymity can jump out of local optimums and improve its solution quality (*i.e.*, increase the number of entry values that are not suppressed) using our Ordered Greed Framework. On average, for all values of k, the percent improvement from using the framework is 34.6%.

Table 4 shows a surprising set of results. More specifically, given that anonymization algorithms often process datasets in column-wise order, we can see here that a random column ordering can significantly lead to higher quality solutions than orderings determined by known greedy heuristics. For example, as age has the most distinct values and the widest range of values (see Table 1), column orders based on the most distinct values [24] or the widest range of values [13] require a leading 0. Notice that a leading 0 only shows up as a best solution in Table 4 one time, namely for $k = 150$. Moreover, a leading 0 appears frequently in the worst solutions. This suggests that these deterministic greedy algorithms from the literature may produce solutions which are far from optimal.

5.3   `OGF` Explores the Search Space Better than `tradGA`

Referring to Figure 14, we again see that `OGF` produces significantly higher quality solutions (*i.e.*, larger number of entry values that are not suppressed) than `tradGA`. Furthermore, as `OGF` produces a high-quality solution very early in the first generation, our framework would be useful for on-demand applications like health research and surveillance, where the user may require results within seconds, or may be willing to wait longer for a potentially higher-quality solution.

When all chromosomes in a population have the same fitness (called *population convergence*) they come to a consensus on the best solution, which is a desirable genetic algorithm feature. Notice that `OGF`'s population converges to a similar resulting fitness, while `tradGA`'s population does not converge and may need more time to explore the search space. This lack of exploration may be due to a larger search space. Consider the element-to-column mapping described in Table 1. `OGF`'s permutation chromosomes represent candidate solutions in a search space size $= 8! = 40320$ for the Adult dataset, whereas `tradGA`'s bit string chromosomes exist in a search space size $= 2^{167-8} = 2^{159} = 7.3 \times 10^{47}$.

5.4   OGF Runs in a Reasonable Amount of Time

From Figure 15, we see that `Datafly` runs very fast for all values of k (3 or 4 seconds). Recall that we expected `Datafly` to run faster than `OGF+Datafly`, as `Datafly` (except for the first four lines) is called as a module many times inside `OGF+Datafly`. Figure 15 also shows that `OGF+Datafly` has a longer runtime than `tradGA` for most values of k. We attribute the longer runtime to more complex fitness evaluations (see Figures 10(b) and 11) and operators (see Figure 12). It is encouraging that at very low k values, `OGF` performs the same or significantly better than `tradGA`; although the choice of k is still not clear, experts in the field say that k need not be large (*e.g.*, 5 or 6 [19], less than 10 [27]).[4] Note that for the smallest k considered here ($k = 5$), the runtime is 71.9 seconds, or 1 minute and 12 seconds. Even in the worst case (*i.e.*, k=200), the runtime is 197 seconds, or 3 minutes and 17 seconds, which is comparable to runtimes of several greedy algorithms in the literature [4,12,26] and is a drastic improvement over Iyengar's 18-hour runtime [8].

A good genetic algorithm needs to balance exploring the search space and exploiting good chromosomes (see Section 2.3). Iyengar [8] may have explored too much with his large population evolving over large number of generations. Given this and knowledge that the fitness function is the most time-consuming aspect of a genetic algorithm for privacy protection, our implementations of `OGF` and `tradGA` use a much smaller population size (10 random chromosomes compared to Iyengar's 5000). To overcome our small population's lack of diversity, we use a high mutation rate (10% compared to Iyengar's 0.2% [8]), which leads to further exploration of the search space.

---

[4] Algorithmics researchers test their algorithms for high values of k (10-1000); however, the Newfoundland and Labrador Centre for Health Information (NLCHI) currently release statistics to the public if $k \geq 5$.

6   Conclusions

The downside to a greedy hill-climbing algorithm is that the solution may only be a local optimum. In this paper, we proposed a new genetic algorithm framework that avoids this inadequacy *and* benefits from greedy optimization. We showed that our Ordered Greed Framework (using `Datafly` as an internal module):

1. produces higher quality solutions (*i.e.*, less suppressed entries) than (a) previous genetic algorithms and (b) `Datafly` alone;
2. allows `Datafly` to jump out of local optimums, causing an average quality improvement of 34.6%;
3. explores the search space better than previous genetic algorithms;
4. runs in a reasonable amount of time.

Also in our investigation of the second point above, we discover a surprising result concerning at least two widely-accepted greedy optimization algorithms in the literature. More specifically, given that anonymization algorithms often process datasets in column-wise order, we show that a random column ordering can significantly lead to higher quality solutions than orderings determined by known greedy heuristics. Furthermore, our implementation of `Datafly` inside the framework includes nearly all extensions listed as future work by leading research in this area [26]: (1) avoiding getting stuck at poor local optimums, (2) suppressing data, (3) local recoding, and (4) generalizing numeric attributes without hierarchies.

Future research includes optimizing our Ordered Greed Framework by systematically exploring parameter settings (*e.g.,* population size, mutation rate, and operators) and other genetic algorithmic techniques (*e.g.,* seeding, elitism, and micro-genetic algorithms). As we intend to eventually offer a working solution for health research and surveillance, we also plan to make minor modifications to our algorithms and adopt techniques from utility research, so that our fitness metric reflects the quality of the data after evaluation. Unless we implement generalization, include metrics for applications like classification [8] and association rule mining [3], and test our implementation on real patient data, it is unlikely that our framework will be used in practice. Finally, it would be interesting to determine if our framework would be useful for other applications that depend on column ordering (*e.g.,* building decision trees).

Acknowledgments

Thank you to David Churchill for his help in codifying the tree data structure. Thanks also to Bradley Malin for his suggestions on additional experimentation and emphasizing the improvement potential of the Ordered Greed Framework. Finally, many thanks for insightful feedback from all anonymous reviewers and workshop organizers.

References

1. Aggarwal, G., Feder, T., Kenthapadi, K., Motwani, R., Panigrahy, R., Thomas, D., Zhu, A.: Anonymizing Tables. In: Eiter, T., Libkin, L. (eds.) ICDT 2005. LNCS, vol. 3363, pp. 246–258. Springer, Heidelberg (2004)
2. Anderson, P., Ashlock, D.: Advances in ordered greed. In: Proc. of 2004 Artificial Neural Networks in Engineering Conf., ANNIE 2004 (2004)
3. Atzori, M., Bonchi, F., Giannotti, F., Pedreschi, D.: Blocking anonymity threats raised by frequent itemset mining. In: Proc. of 5th IEEE Int'l Conf. on Data Mining, ICDM 2005, pp. 561–564 (2005)
4. Bayardo, R., Agrawal, R.: Data privacy through optimal k-anonymization. In: Proc. of 21st IEEE Int'l Conf. on Data Engineering, ICDE 2005, pp. 217–228 (2005)
5. Byun, J., Kamra, A., Bertino, E., Li, N.: Efficient k-Anonymization using clustering techniques. In: Kotagiri, R., Radha Krishna, P., Mohania, M., Nantajeewarawat, E. (eds.) DASFAA 2007. LNCS, vol. 4443, pp. 188–200. Springer, Heidelberg (2007)
6. Chaytor, R.: Utility Preserving k-Anonymity. Technical report MUN-CS 2006-01, Dept. Computer Science, Memorial University (2006)
7. Fung, B., Wang, K., Yu, P.: Top-down specialization for information and privacy preservation. In: Proc. of 21st IEEE Int'l Conf. on Data Engineering, ICDE 2005, pp. 205–216 (2005)
8. Iyengar, V.: Transforming data to satisfy privacy constraints. In: Proc. of 8th ACM SIGKDD Int'l Conf. on Knowledge Discovery and Data Mining, KDD 2002, pp. 279–288 (2002)
9. Jiang, W., Clifton, C.: Privacy-preserving distributed k-anonymity. In: Proc. of 19th IFIP WG 11.3 Working Conf. on Data and Applications Security, DBSec 2005, pp. 166–177 (2005)
10. Kifer, D., Gehrke, J.: Injecting utility into anonymized datasets. In: Proc. of 2006 ACM SIGMOD Int'l Conf. on Management of Data, pp. 217–228 (2006)
11. Koudas, N., Srivastava, D., Yu, T., Zhang, Q.: Aggregate query answering on anonymized tables. In: Proc. of 23rd IEEE Int'l Conf. on Data Engineering, ICDE 2007 (2007)
12. LeFevre, K., DeWitt, D., Ramakrishnan, R.: Incognito: Efficient full-domain k-anonymity. In: Proc. of 2005 ACM SIGMOD Int'l Conf. on Management of Data, pp. 49–60 (2005)
13. LeFevre, K., DeWitt, D., Ramakrishnan, R.: Mondrian multidimensional k-anonymity. In: Proc. of 22nd IEEE Int'l Conf. on Data Engineering, ICDE 2006 (2006)
14. LeFevre, K., DeWitt, D., Ramakrishnan, R.: Workload-aware anonymization. In: Proc. of 12th ACM SIGKDD Int'l Conf. on Knowledge Discovery and Data Mining, KDD 2006, pp. 277–286 (2006)
15. Li, N., Li, T., Venkatasubramanian, S.: t-closeness: privacy beyond k-anonymity and ℓ-diversity. In: Proc. of IEEE 23rd Int'l Conf. on Data Engineering, ICDE 2007 (2007)
16. Lin, X., Kwok, Y., Lau, V.: A genetic algorithm based approach to route selection and capacity flow assignments. Computer Communications 26(9), 961–974 (2003)
17. Lunacek, M., Whitley, D., Ray, I.: A crossover operator for the k-anonymity problem. In: Proc. of 7th ACM Genetic and Evolutionary Computation Conf., GECCO 2006, pp. 1713–1720 (2006)
18. Machanavajjhala, A., Kifer, D., Gehrke, J., Venkitasubramaniam, M.: ℓ-diversity: privacy beyond k-anonymity. In: Proc. of 22nd IEEE Int'l Conf. on Data Engineering, ICDE 2006 (2006)

19. Meyerson, A., Williams, R.: On the complexity of optimal k-anonymity. In: Proc. of 23rd ACM Sym. on Principles of Database Systems, PODS 2004, pp. 223–228 (2004)
20. Mitchell, M.: An Introduction to Genetic Algorithms. MIT Press, Cambridge (1999)
21. Pandey, R., Chattopadhyay, S.: Low power technology mapping for LUT based FPGA - a genetic algorithm approach. In: Proc. of 16th IEEE Int'l Conf. on VLSI Design, VLSI 2003, pp. 79–84 (2003)
22. Samarati, P., Sweeney, L.: Protecting Privacy when Disclosing Information: k-Anonymity and its Enforcement through Generalization and Suppression. Technical report SRI-CSL-98-04. SRI International, Computer Science Laboratoy (1998)
23. Sweeney, L.: Guaranteeing anonymity when sharing medical data, the datafly system. In: Conf. of American Medical Informatics Association, Annual Fall Sym., AMIA 1997, pp. 51–55 (1997)
24. Sweeney, L.: Achieving k-anonymity privacy protection using generalization and suppression. Int'l J. on Uncertainty, Fuzziness and Knowledge-Based Systems 10(5), 571–588 (2002)
25. Truta, T., Vinay, B.: Privacy protection: p-sensitive k-anonymity property. In: Proc. of 22nd IEEE Int'l Conf. on Data Engineering Workshops (2006)
26. Wang, K., Yu, P., Chakraborty, S.: Bottom-up generalization: a data mining solution to privacy protection. In: Perner, P. (ed.) ICDM 2004. LNCS (LNAI), vol. 3275, pp. 249–256. Springer, Heidelberg (2004)
27. Winkler, W.: Using Simulated Annealing for k-Anonymity. Technical report 2002-07. U.S. Bureau of the Census, Statistical Research Division (2002)
28. Wong, R., Li, J., Fu, A., Wang, K. (α, k)-anonymity: an enhanced k-anonymity model for privacy preserving data publishing. In: Proc. of 12th ACM SIGKDD Int'l Conf. on Knowledge Discovery and Data Mining, KDD 2006, pp. 754–759 (2006)
29. Xiao, X., Tao, Y.: Personalized privacy preservation. In: Proc. of 2006 ACM SIGMOD Int'l Conf. on Management of Data, pp. 229–240 (2006)
30. Xu, J., Wang, W., Pei, J., Wang, X., Shi, B., Fu, A.: Utility-based anonymization for privacy preservation with less information loss. ACM SIGKDD Explorations 8(2), 21–30 (2006)

Probabilistic Anonymity

Sachin Lodha[1] and Dilys Thomas[2],*

[1] Tata Research Development and Design Centre, Pune, India
[2] Google Inc., Mountain View, USA
sachin.lodha@tcs.com, dilysthomas@google.com

Abstract. In this age of globalization, organizations need to publish their micro-data owing to legal directives or share it with business associates in order to remain competitive. This puts personal privacy at risk. To surmount this risk, attributes that clearly identify individuals, such as Name, Social Security Number, and Driving License Number, are generally removed or replaced by random values. But this may not be enough because such de-identified databases can sometimes be joined with other public databases on attributes such as Gender, Date of Birth, and Zipcode to re-identify individuals who were supposed to remain anonymous. In the literature, such an identity-leaking attribute combination is called as a quasi-identifier. It is always critical to be able to recognize quasi-identifiers and to apply to them appropriate protective measures to mitigate the identity disclosure risk posed by join attacks.

In this paper, we start out by providing the first formal characterization and a practical technique to identify quasi-identifiers. We show an interesting connection between whether a set of columns forms a quasi-identifier and the number of distinct values assumed by the combination of the columns. We then use this characterization to come up with a probabilistic notion of anonymity. Again we show an interesting connection between the number of distinct values taken by a combination of columns and the anonymity it can offer. This allows us to find an ideal amount of generalization or suppression to apply to different columns in order to achieve probabilistic anonymity. We work through many examples and show that our analysis can be used to make a published database conform to privacy rules like HIPAA. In order to achieve probabilistic anonymity, we observe that one needs to solve multiple 1-dimensional k-anonymity problems. We propose many efficient and scalable algorithms for achieving 1-dimensional anonymity. Our algorithms are optimal in a sense that they minimally distort data and retain much of its utility.

1 Introduction

"Over a year and a half, one individual impersonated me to procure over $50,000 in goods and services. Not only did she damage my credit, but she escalated her crimes to

* Supported in part by NSF Grant ITR-0331640. This work was also supported in part by TRUST (The Team for Research in Ubiquitous Secure Technology), which receives support from the National Science Foundation (NSF award number CCF-0424422) and the Air Force Office of Scientific Research (AFOSR award number 021032-003). This work was initiated when the second author was a summer intern at Tata Research Development and Design Centre, Pune, India.

F. Bonchi et al. (Eds.): PinKDD 2007, LNCS 4890, pp. 56–79, 2008.

a level that I never truly expected: she engaged in drug trafficking. The crime resulted in my erroneous arrest record, a warrant out for my arrest, and eventually, a prison record when she was booked under my name as an inmate in the Chicago Federal Prison." - An excerpt from the verbal testimony of Michelle Brown to a US Senate Committee [11].

Unfortunately, in today's highly networked digital world, incidents like the above with Michelle Brown are commonplace. According to the Bureau of Justice Statistics Bulletin [8], 3.6 million households, representing 3% of the households in the United States, discovered that at least one member of the household had been the victim of identity theft during the previous 6 months in 2004. According to the same report, the estimated loss as a result of identity theft was about $ 3.2 billion. Needless to say that preventing identity thefts is one of the top priorities for government, corporations and society alike.

Globalization further complicates this picture. Due to legal directives or business associations, there are multiple scenarios where organizations need to share or publish their micro-data to remain competitive. This puts personal privacy at further risk. To surmount this risk, attributes that clearly identify individuals, such as Name, Social Security Number, Driving License Number, are generally removed or replaced by random values. But this may not be enough because such de-identified databases can sometimes be joined with other public databases on seemingly innocuous attributes to re-identify individuals who were supposed to remain anonymous. For example, according to one study [33], approximately 87% of the population of the United States can be uniquely identified on the basis of Gender, Date of Birth, and 5-digit Zipcode. The uniqueness of such attribute combinations leads to a class of attacks where data is re-identified by joining multiple and often publicly available datasets. This type of attack was illustrated by Sweeney in [33] where the author was able to join a public voter registration list and the de-identified patient data of Massachusetts' state employees to determine the medical history of the state's governor.

In literature, such an identity-leaking attribute combination is called as a *quasi-identifier*. It is always critical to be able to recognize quasi-identifiers and to apply appropriate protective measures to mitigate the identity disclosure risk posed by join attacks. In fact, Sweeney herself proposed a k-anonymity model in [35] for the same. According to her, a database table is said to be k-anonymous if for each row in the table there are $k - 1$ other rows in the table that are *identical* along the quasi-identifier attributes. Clearly, a join with a k-anonymous table would give rise to k or more matches and create confusion. Thus, an individual is hidden in a crowd of size k giving her k-anonymity. It also means that the identity disclosure risk is at most $1/k$ for the "join" class of attacks.

Although such a simple and clear quantification of privacy risk makes the k-anonymity model attractive, its widespread use in practice is severely hampered owing to the following factors:

1. Choice of k is not clear. From a pure privacy point of view, larger k would mean more privacy, but it comes at the cost of utility [3]. What is the right choice of k for the given data and the given notion of utility has not been very well understood yet.
2. For the k-anonymity model to be effective, it is critical that there is a complete understanding of the quasi-identifiers for the given data-set. But there is no real

formalism available for deciding whether an attribute combination could form a quasi-identifier. This is currently done manually, based on folk-lore and human expertise.

3. For a given k, the goal is always to minimally suppress or generalize the data such that the resultant data-set is k-anonymous. However, for some natural notions of measuring this resultant distortion, the minimization problems turn out to be NP-Hard [26, 4, 6].

On the approximation front, no efficient but good approximation algorithms are currently known. The known algorithms are either $\tilde{O}(k)$ approximations [26, 4] or super-linear [6] - thus making them inefficient or expensive.

1.1   Paper Organization and Contribution

In this paper, we start out by providing the first formal characterization and a practical technique to identify quasi-identifiers. In Section 2, we also show an interesting connection between whether a set of columns forms a quasi-identifier and the number of distinct values assumed by the combination of the columns.

We then use this characterization in Section 3 to come up with a probabilistic notion of anonymity. Again we show an interesting connection between the number of distinct values taken by a combination of columns and the anonymity it can offer. This allows us to find an ideal amount of generalization or suppression to apply to different columns in order to achieve probabilistic anonymity. We work through many examples and show that our analysis can be used to make a published database conform to privacy rules like HIPAA.

In order to achieve the probabilistic anonymity, we observe that one needs to solve multiple 1-dimensional k-anonymity problems. In Section 4, we propose many efficient and scalable algorithms for achieving 1-dimensional anonymity. Our algorithms are optimal in a sense that they minimally distort data and retain much of its utility. The algorithms provided are a stark contrast to previous NP-hard results and comparatively more complicated algorithms for the previous notion of anonymity called k-anonymity [35].

We then experimentally verify our algorithms on real life data sets in Section 5. We sketch the related work in Section 6 and finally conclude in Section 7.

2   Automatic Detection of Quasi-identifiers

Definition 1. *A quasi-identifier set Q is a minimal set of attributes in table T that can be joined with external information to re-identify individual records (with sufficiently high probability).*

Above definition is from [29]. A similar definition can be found in an earlier paper of Dalenius [16]. As the reader can sense, this definition is informal since it does not make "external information" and "sufficiently high probability" explicit. Possibly because of this, we do not know any formal procedure or test for identifying quasi-identifiers. Almost always, researchers and practitioners assume that quasi-identifier attribute sets are known based on a specific knowledge domain [23].

We present a more formal definition of quasi-identifier below. In our definition, we do not insist on minimality of attribute set although one could easily accommodate it if required. The external information is the *universal table* $\mathcal{U}$ having information about an entire (relevant) population. It has n rows. Typically, $\mathcal{U}$ would mean census records that many countries make readily available [2]. The columns of the universal table include the quasi-deintifier columns among other columns.

Definition 2. α**-quasi-identifier.** *An α quasi-identifier is a set of attributes along which an α fraction of rows in the universe can be uniquely identified by values along the combination of these attribute columns.*

Example 1. *Empirically it has been observed that 87% of the people in the U.S. can be uniquely identified by the combination of* Gender, Date of Birth *and* Zipcode. *Therefore (*Gender, Date of Birth, Zipcode*) forms a 0.87-quasi-identifier for the U.S. population. Note that the U.S. census table is our universal table $\mathcal{U}$ here.*

Ideally, given an α and $\mathcal{U}$, it is straight-forward to figure out whether some particular attribute combination forms an α-quasi-identifier in $\mathcal{U}$ by simply measuring the number of singletons in that attribute combination. One may even try an apriori like approach [7] and calculate all α-quasi-identifiers in $\mathcal{U}$. In practice, there are errors in $\mathcal{U}$ that come in during data collection phase itself [12, 1] and the knowledge about $\mathcal{U}$ is never exact. This would lead to erroneous conclusions about a quasi-identifier. Therefore, it does not justify the expensive calculations given above. In fact, one then prefers a quick and inexpensive approach that gives a good *estimate* of the same.

In what follows, we assume that the universal table $\mathcal{U}$ itself is not known. What we know is that it is a *random sample* built *with replacement* from a probability space. Thus our analysis is probabilistic. For the sake of analysis, we require that there is a probability distribution, but in reality, our final results are independent of this probability distribution. Moreover, we work only with the expectations since our goal is to give *good estimates* quickly. Since the sum of random variables is tightly concentrated around the expectation (by bounds like the Chernoff bounds [15]), our analysis and results are quite fair. We do not work out the Chernoff analysis though in order to keep our results and presentation simple.

We build our probability space on the distinct values that an attribute combination can take. Therefore, we need to know the number of distinct values for every attribute combination. Since one can get (or reasonably estimate) the count of distinct values for each attribute in $\mathcal{U}$ [17], we simplify our task with the following assumption.

Definition 3. Multiple Domain Assumption. *Let d_1, d_2, ..., d_k be the number of distinct values along columns C_1, C_2, ..., C_k respectively. Then, the total number of distinct values taken by the $(C_1, C_2, \ldots, C_k)$ column set is $D = d_1 \times d_2 \times \ldots d_k$.*

Example 2. *We study the number of distinct values taken by the set of columns (*Gender, Date of Birth, Zipcode*). The number of distinct values of column* Gender *(C_1) is $d_1 = 2$. The number of distinct values of column* Date of Birth *(C_2) can be approximated as $d_2 = 60 * 365 \approx 2 * 10^4$.[1] The number of distinct values*

---

[1] Throughout this paper we assume that the ages of people belonging to the database comes from an interval of size 60 years.

along column Zipcode *(C_3) is $d_3 = 10^5$. The number of distinct values of the column-set* (Gender, Date of Birth, Zipcode) *is $D = d_1 \times d_2 \times d_3 = 2 * (2 * 10^4) * 10^5 = 4 * 10^9$.*

As another example, consider the set of columns (Nationality, Date of Birth, Occupation). *The number of distinct values of column* Nationality *(C_1) is $d_1 = 200$. Once again, the number of distinct values of column* Date of Birth *(C_2) can be approximated as $d_2 = 60 * 365 \approx 2 * 10^4$. The number of distinct values of column* Occupation *(C_3) is roughly $d_3 = 100$. Thus $D = d_1 \times d_2 \times d_3 = 200 * (2 * 10^4) * 100 = 4 * 10^8$.*

Remark. The multiple domain assumption is a weak bound especially is the columns are correlated. Please note that it may be possible to consider correlations among various attributes and, therefore, arrive at a tighter estimate of D. Such analysis would certainly lead to improved bounds in what follows. Yet we decided not to incorporate correlations - partly because it would have made analysis very tough and the main purport of our results could have easily been lost, but largely because we also wanted our results to be viable and useful. Readers will notice that larger estimate for D implies stricter privacy control and more anonymization in what follows. This is acceptable in practice as long as it is easily doable and does not lead to high loss in data utility.

Suppose that a set of columns take D different values with probabilities $p_1, p_2, \ldots, p_D$, where $\sum_{i=1}^{D} p_i = 1$. Let us first calculate the probability that the ith element is a singleton in the universal table $\mathcal{U}$. It means first selecting one of the entries in the table (there are n choices), setting it to be this i^{th} element (which has probability p_i), and setting all other entries in the table to something else (which happens with probability $(1 - p_i)^{n-1}$). Thus, the probability of ith element being a singleton in the universal table $\mathcal{U}$ is $np_i(1 - p_i)^{n-1}$.

Let X_i be the indicator variable representing whether the ith element is a singleton. Then, its expectation

$$E[X_i] = P[X_i = 1] = np_i(1 - p_i)^{n-1} \approx np_i e^{-np_i}.$$

Let $X = \sum_{i=1}^{D} X_i$ be the counter for the number of singletons. Now its expectation is given by

$$E[X] = \sum_{i=1}^{D} E[X_i] = \sum_{i=1}^{D} np_i e^{-np_i}.$$

Let us analyze which distribution maximizes this expected number of singletons. We aim to maximize $\sum_{i=1}^{D} x_i e^{-x_i}$, subject to $\sum_{i=1}^{D} x_i = n$ and $0 \leq x_i, \forall 1 \leq i \leq D$.

Theorem 1. *If $D \leq n$, then the expected number of singletons is bounded above by $\frac{D}{e}$.*

PROOF.[of theorem 1] If $f(x) = xe^{-x}$, $f'(x) = (1 - x)e^{-x}$ and $f''(x) = (x - 2)e^{-x}$. Thus, the function f has a global maximum at $x = 1$, since $f'(1) = 0$ and $f''(1) < 0$.

Now the expected number of singletons,

$$\sum_{i=1}^{D} x_i e^{-x_i} \leq \sum_{i=1}^{D} e^{-1} = \frac{D}{e}.$$

This expression is a tight upper bound on the expected number of singletons for $D \leq n$. For example, it is almost obtained by setting $x_i = 1$, for $i = 1, 2, \ldots, D - 1$, and $x_D = n - D + 1$. □

Theorem 2. *If $D \geq n$, then the expected number of singletons is bounded above by $ne^{\frac{-n}{D}}$.*

PROOF.[of theorem 2] If $f(x) = xe^{-x}$, $f'(x) = (1 - x)e^{-x}$ and $f''(x) = (x - 2)e^{-x}$. The function f has a point of inflection at $x = 2$, since $f''(x) < 0$ for $x < 2$ implying the function is concave here, and $f''(x) > 0$ for $x > 2$ implying the function is convex here.

First we claim that on maximizing $\sum_{i=1}^{D} x_i e^{-x_i}$, no $x_i \geq 2$. Suppose otherwise: after maximizing $\sum_{i=1}^{D} x_i e^{-x_i}$, some $x_a \geq 2$. As $D \geq n$, and $\sum_{i=1}^{D} x_i = n$, some $x_b < 1$. For some small δ, replacing x_a by $x_a - \delta$ and x_b by $x_b + \delta$ we retain $\sum_{i=1}^{D} x_i = n$. As $f(x) = xe^{-x}$ increases towards x=1, $f(x_a - \delta) > f(x_a)$ and $f(x_b + \delta) > f(x_b)$. Thus $\sum_{i=1}^{D} x_i e^{-x_i}$ is increased, contradicting the fact that it was maximized. Thus, $\forall 1 \leq i \leq D$, $x_i < 2$.

Now $f''(x) < 0$ for $0 \leq x < 2$. Since f is concave, we can apply Jensen's inequality [28] [2] to get

$$\sum_{i=1}^{D} x_i e^{-x_i} = D \sum_{i=1}^{D} \frac{1}{D} x_i e^{-x_i}$$

$$\leq D \cdot (\sum_{i=1}^{D} \frac{x_i}{D}) e^{-(\sum_{i=1}^{D} \frac{x_i}{D})}$$

$$= ne^{\frac{-n}{D}}.$$

Thus, if $D \geq n$, the expected number of singletons is bounded above by $ne^{\frac{-n}{D}}$. □

Figure 1 shows how the maximum expected fraction of singletons or unique rows in a collection of n rows behaves, as the number of distinct values, D, varies. The graph plots the maximum expected fraction of unique rows as a function of $\frac{D}{n}$. It is the line $\frac{D}{en}$ for $\frac{D}{n} \leq 1$ according to Theorem 1. For $\frac{D}{n} \geq 1$, it is the curve $e^{\frac{-n}{D}}$ according to Theorem 2. The curve is both continuous and smooth (differentiable) at $\frac{D}{n} = 1$ with $f(1) = \frac{1}{e}$ and $f'(1) = \frac{1}{e}$.

Figure 1 forms a ready reference table in order to test whether a set of attributes forms a probable quasi-identifier. For example, if for a set of attributes $D < 3n$, then it is unlikely that the set of attributes will form a 0.75 quasi-identifier. If a set of attributes do not form an α-quasi-identifier according to the the number of distinct values in Figure 1, then they almost certainly do not form an α-quasi-identifier as the plot gives the maximum expected fraction of singletons (as per Theorem 1 and Theorem 2).

Example 3. *We now show how (Gender, Date of Birth, Zipcode) forms a quasi-identifier when restricted to the U.S. population. The size of the U.S. population can be approximated as $3 * 10^8$, that is, the size of the universal table n is $3 * 10^8$. The number of distinct values taken by the attribute set (Gender, Date of Birth,*

---

[2] If f is a concave function, and $\sum_{i=1}^{m} p_i = 1$, with $p_i \geq 0$ $\forall i$, then $\sum_{i=1}^{m} p_i f(x_i) \leq f(\sum_{i=1}^{m} p_i x_i)$.

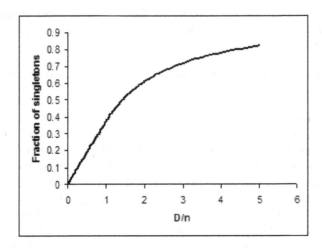

Fig. 1. Quasi-Identifier Test

$\mathtt{Zipcode})$ *is* $4 * 10^9$ *from Example 2. Therefore, by Theorem 2, the maximum expected fraction of rows with singleton occurrence is* $e^{-3*10^8/4*10^9} = e^{-0.075} \approx 0.93$. *Thus,* $(\mathtt{Gender},\ \mathtt{Date}\ \mathtt{of}\ \mathtt{Birth},\ \mathtt{Zipcode})$ *is a potential* 0.93 *quasi-identifier. Please recall that this combination is already known to be a* 0.87 *quasi-identifier [35].*

Example 4. *We now give an example of a set of attributes that does not form a quasi-identifier. Let us consider* $(\mathtt{Nationality},\ \mathtt{Date}\ \mathtt{of}\ \mathtt{Birth},\ \mathtt{Occupation})$. *The number of distinct values along these columns is given from Example 2 as* $D = 4 * 10^8$. *Here the size of the universal table is* $n = 6 * 10^9$, *that is, equal to the world population. Since* $D < n$, *we use Theorem 1 and find that the expected fraction of rows with singleton occurrence is bounded above by* $D/en = 4 * 10^8/2.7 * 6 * 10^9 \approx 0.025$. *Thus these columns almost certainly do not form even a* 0.05 *quasi-identifier as* 0.025 *is an upper bound on the expected fraction of singletons over all possible probability distributions over quasi-identifier values.*

We now provide a simple test to decide whether a combination of attributes forms a potentially dangerous quasi-identifier, that is, say $\alpha \geq 0.5$.

Theorem 3. *Given a universe of size n, a set of attributes can form an α-quasi-identifier (where $0.5 \leq \alpha < 1$) if the number of distinct values along the columns,* $D > \frac{n}{ln(1/\alpha)}$.

Proof (of theorem 3). Note that $D > n$. If not, then, by Theorem 1, the maximum expected fraction of rows taking unique values is $D/en \leq 1/e < \alpha$.

From Theorem 2, the maximum expected fraction of rows taking unique values along the columns with D distinct values is $e^{-n/D}$. For the the set of rows to form an α-quasi-identifier, this fraction must be larger than α. Thus, $e^{-n/D} > \alpha$, which implies that $D > \frac{n}{ln(1/\alpha)}$.

2.1   Distinct Values and Quasi-identifiers

In this section, we have provided an interesting connection between whether a set of columns forms a quasi-identifier and the number of distinct values assumed by the combination of the columns. The main contributions of this association are as follows.

1. We provide a fast and efficient technique to test whether a set of columns forms a quasi-identifier. However there may be false positives. A set of columns signalled as a probable α quasi-identifier may only be a β quasi-identifier for some $\beta < \alpha$.
2. We do not assume anything about the distribution on the values taken by the quasi-identifier. The expected number of singletons is bounded by the expression provided in this section for all possible distributions over the values taken by the quasi-identifier.
3. When a set of columns is declared not to be a quasi-identifier by the test in this section, the set of columns is almost certainly not a quasi-identifier, that is, there is a minuscule chance of false negatives.

3   Probabilistic Anonymity

In Sweeney's anonymity model [35], every row of the dataset is required to be identical with k other rows in the dataset along Q. In the following notion of anonymity, we insist that each row of the anonymized dataset should match with at least k or more rows of the universal table $\mathcal{U}$ along Q. Since $\mathcal{U}$ is represented in a probabilistic fashion, we want this event to happen with high probability.

Definition 4. *A dataset is said to be probabilistically $(1 - \beta, k)$- anonymized along a quasi-identifier set Q, if each row matches with at least k rows in the universal table $\mathcal{U}$ along Q with probability greater than $(1 - \beta)$.*

Our notion of anonymity is similar to that of [35] for an adversary who is *oblivious*, that is, she is not really looking for some *particular* individuals, but is trying to do a join on Q and checking if she is "lucky". This kind of attack is quite a possibility in today's outsourcing scenarios where an attacker, say, from a call center, would want to know identities in her client's data without really knowing whom to look for. If an adversary is looking for a *particular* individual in the anonymized dataset, then Sweeney's model (k-map [34]) would generally provide better privacy than our model for it would always yield k matches. For our model to work well against such an adversary, we need to declare the original dataset itself as the universal table $\mathcal{U}$ and carry out anonymization. Note that Sweeney's original model of k-anonymity is oblivious of the universal table $\mathcal{U}$.

  In what follows, we build on the strong connection between the number of distinct values assumed by a set of attributes Q and its identity revealing potential that was discovered in Section 2. Intuitively, it is clear from Theorems 1, 2 and 3 that the potency of Q as a quasi-identifier would decrease if we reduce the number of distinct values assumed by Q. This is to be done with appropriate *generalization*. We borrow the following definition of generalization from [35] which has an excellent discussion on this topic.

Definition 5. *Generalization involves replacing (or recoding) a value with a less specific but semantically consistent value.*

Example 5. *The original ZIP codes $\{02138, 02139\}$ can be generalized to $0213*$, thereby stripping the rightmost digit and semantically indicating a larger geographical area.*

One way of looking at generalization is creating $<< D$ partitions of the space of D distinct values and choosing a representative for each partition. In fact, it would give us k-anonymity if we could ensure that most of these partitions are represented by k or more of their own members in the universal table $\mathcal{U}$ with high probability. To make this work, let us suppose that we have a D'-partition of original D size space such that each partition has probability $1/D'$ (or $O(1/D')$ to be precise). Given a $< p_1, p_2, \ldots, p_D >$ probabilities of the original D size space, such partitioning is certainly possible using techniques we show in Section 4 for a single dimension. Now, we analyze below the bound on D' that is necessary in order to ensure that most of these partitions are represented k or more times in $\mathcal{U}$ with high probability. Please recall that $\mathcal{U}$ has size n and it is built by sampling with replacement.

Theorem 4. *A data set is probabilistically $(1 - \beta, k)$-anonymized with respect to a universal table $\mathcal{U}$ of size n along the quasi-identifier Q if the number of distinct values along Q, $D' < \frac{n}{k}(1 - c)$ for some small constant c.*

Before we proceed with the proof, please note that Theorem 4 provides a recommendation for D', the number of partitions of D size space of Q. If the probabilities $< p_1, p_2, \ldots, p_D >$ are known, then as per our earlier assumption, one could cluster these probabilities such that D' equi-probable partitions are created. This concretizes generalization which could be used by any data-holder for anonymizing its data before release.

Proof (of theorem 4). Let us suppose that we have got a D'-partition of original D size space of quasi-identifier Q such that each partition has probability $1/D'$. Let X_i denote the indicator variable if $\geq k$ rows in the universal table $\mathcal{U}$ are chosen from the ith partition.

$$P[X_i = 1] = \sum_{j=k}^{n} \binom{n}{j}(\frac{1}{D'})^j(1 - \frac{1}{D'})^{n-j}$$

$$= 1 - \sum_{j=0}^{k-1} \binom{n}{j}(\frac{1}{D'})^j(1 - \frac{1}{D'})^{n-j}$$

$$\geq 1 - exp(\frac{-D'(n/D' - (k-1))^2}{2n})$$
$$\text{(by Chernoff bounds [15])}$$

$$= 1 - exp(\frac{-(n - (k-1)D')^2}{2nD'}).$$

For $1 - \beta$ probability guarantee, we would like to have

$$1 - exp(\frac{-(n - (k - 1)D')^2}{2nD'}) \geq 1 - \beta,$$

that is,

$$\frac{-(n - (k - 1)D')^2}{2nD'} \leq ln\beta.$$

This is true when,

$$0 \leq D'^2 + \frac{2nD'}{k - 1}\left(\frac{ln\beta}{k - 1} - 1\right) + \left(\frac{n}{k - 1}\right)^2,$$

that is,

$$D' \leq \frac{n}{k - 1}(1 + x - \sqrt{x^2 + 2x}),$$

where

$$x = \frac{-ln\beta}{k - 1}.$$

This implies that

$$D' \leq \frac{n}{k}(1 - c)$$

is sufficient for some small constant c.

Example 6. *Let $\mathcal{U}$ be the U.S. Census Table of size $n = 3 * 10^8$. Consider the columns $Q = $ (Gender, Date of Birth, Zipcode). By Example 2, $D = 4 * 10^9$. According to Theorem 4, a dataset is $(0.9, 100)$ anonymized along Q with respect to $\mathcal{U}$ if we make D' partitions (or generalizations) of the D size space where $(n/125 < n/100 * (1 - c))$*

$$D' \leq \frac{n}{125} = 2.4 * 10^6.$$

Thus, we have to reduce the number of possibilities for Q by a factor of $D/D' < 1700$. Consider the following generalization (Gender, Half-year of Birth, First Four Digits of Zipcode). Now $D' = d'_1 * d'_2 * d'_3$. d'_1, the number of distinct values of Gender, is 2. d'_2 is $60 * 2 = 120$, and $d'_3 = 10^4$. Therefore, $D' = 2.4 * 10^6$. This should be good enough to make each row 100-anonymous with probability at least 0.9.

3.1 Privacy vs Utility

Note that (Gender, Half-year of Birth, First Four Digits of Zipcode) was just one of many different ways we could have compressed the D size space in Example 6 by factor 1700. Ideally, we would like to devise this generalization such that there is little or no loss in the *data utility*. We frame this problem as an optimization problem below where the goal is to retain maximum utility given privacy constraints.

Let there be m columns $< C_1, C_2, \ldots, C_m >$ that need generalization and $w_1, w_2, \ldots, w_m$ be their respective weights giving their relative importance [31, 32]. We aim to anonymize this multi-column database so that maximum utility is retained in the probabilistically k-anonymized output.

Let $d_1', d_2', \ldots, d_m'$ be the number of distinct values along columns $C_1, C_2, \ldots, C_m$ after probabilistic k-anonymization. Then, by Theorem 4,

$$\prod_{i=1}^{m} d_i' = \frac{n}{k}(1 - c) = D'.$$

Let us suppose that the quantile based anonymization from Section 4 is used. Thus, d_i' different quantiles are used along the column C_i. Then, the rank difference of the transformation (from Section 4) is approximately $(\frac{n}{d_i'})^2 \times d_i' = \frac{n^2}{d_i'}$.

The sum of the distortion along all columns weighted by the column weights is, therefore, $n^2(\sum_{i=1}^{m} \frac{w_i}{d_i'})$. Minimizing this is equivalent to minimizing $\sum_{i=1}^{m} \frac{w_i}{d_i'}$ subject to $\prod_{i=1}^{m} d_i' = D'$. For a fixed value of product, the sum of numbers is minimized when all the numbers are equal. Therefore,

$$\frac{w_1}{d_1'} = \frac{w_2}{d_2'} = \ldots \frac{w_m}{d_m'} = \frac{1}{d} \text{ (say)}.$$

Therefore, $d_i' = d \times w_i \ \forall 1 \le i \le m$. The product condition implies, $\prod_{i=1}^{m} d_i' = d^m \prod_{i=1}^{m} w_i = D'$. Therefore,

$$d = \left(\frac{D'}{\prod_{i=1}^{m} w_i}\right)^{1/m},$$

$$d_i' = \left(\frac{D'}{\prod_{i=1}^{m} w_i}\right)^{1/m} \times w_i. \tag{1}$$

Note that if d_i' is less than the number of distinct values in column i initially, say d_i, it suggests applying an approach like quantiles proposed here on column C_i. If d_i' is greater than the number of distinct values in column C_i initially, say d_i, then the column C_i is left untouched. The number of distinct elements for other columns can be recalculated (and increased) after this. That is, if $d_i' > d_i$, then the optimization problem over all other variables is first solved after column C_i is eliminated, i.e. Maximize $\sum_{j=1, j \ne i}^{m} \frac{w_j}{d_j'}$ subject to $\prod_{j=1, j \ne i}^{m} d_j' = D'/d_i$.

Example 7. *Suppose that we want to probabilistically* $(0.9, 100)$-*anonymize a dataset with 3 columns (*Gender, Date of Birth, Zipcode*) and all columns are equally important, that is , they have equal weight.*

As worked out in Example 6, each row is given 100-anonymity with probability at least 0.9 if $D' = 2.4 * 10^6$. As all 3 columns have equal weight, we get $d_1' = d_2' = d_3' \approx 133$. However Gender has only $2 < d_1'$ values. This means we have to leave it untouched and work with the remaining two attributes. That gives $d_2' * d_3' = 1.2 * 10^6$. Since both the columns have equal weight, we get $d_2' = d_3' \approx 1.1 * 10^3$. As $d_2' = 1.1 * 10^3$ is approximately 60 (years)*12 (number of months per year), Date of Birth *is approximated to the month of birth. Also the number of distinct values of* Zipcode *being $O(10^3)$ implies that the last two digits of* Zipcode *are starred out. Thus the anonymization produced is (*Gender, Month of Birth, First Three Digits of Zipcode*).*

Note that this anonymization was entirely worked out in constant time in the above example. For general case, where the number of columns is m, it would require $O(m^2)$ time. Previous techniques to provide anonymity were not only NP-hard in the input size (that means it took exponential time in the dataset) [26,5] but even approximations required many passes over the database [5,6]. [23] required passes to be exponential in the number of columns to be anonymized as the lattice developed there took exponential time to be built.

Example 8. *According to HIPAA [19], each person must be anonymized in a crowd of $k = 20,000 = 2 * 10^4$ people. Now, suppose we want to anonymize a medical records table with columns (Gender, Age (In Years), Zipcode, Disease).*

As always, the U.S. Census Table is the universal table $\mathcal{U}$ with $n = 3 * 10^8$ rows. The quasi-identifier is (Gender, Age (In Years), Zipcode). As the number of distinct values of Gender and Age are 2 and 100 respectively, the number of distinct values of Zipcode allowed is approximately $3 * 10^8/((2 * 10^4) * 2 * 100) = 75$ by Theorem 4. Therefore, Zipcode must be anonymized to its first two digits and should only indicate the State.

3.2 The Curse of Dimensionality

As the number of dimensions (columns) increase, the number of distinct values per column on anonymization decrease rapidly. For example, consider a database table with 25 columns. The aim is to anonymize the table so that 10-anonymity is achieved for the U.S. population of size $3 * 10^8$. Further suppose that all the columns are given equal weight (importance). Applying Theorem 4 and the Multiple Domain Assumption, the number of distinct values per column can be obtained to be roughly 2. Thus all values in a column are generalized to two intervals or converted to two types of values. This hints at reduced data utility measured by any reasonable metric.

This phenomenon was also observed as the curse of dimensionality on k-anonymity [3]. However, we must notice that the previous analysis should only be applied to columns that are available publicly. For example, in the Adults database [10], columns capgain, caploss, fnlwgt and income can be assumed to be sensitive columns that are present only in the database itself and are not available for an external join.

3.3 Distinct Values and Anonymity

In this section, we have provided an interesting connection between the number of distinct values taken by a combination of columns and the anonymity it can offer. The main contributions of this association are as follows.

1. This association between distinct values and anonymity guarantee results in an easy technique to obtain a k-anonymized dataset. Merge similar distinct values taken by a column so that the number of distinct values assumed by the column is reduced. The appropriate reduction in the number of distinct values leads to the conversion of a quasi-identifier into k-anonymous columns. As explained in Section 3.1, this would also help retain much of data utility since it minimally distorts ranks. We shall discuss this angle in more detail in the next section.

2. It also helps in coming up with the right kind of generalization for publicly known attributes so that published database can conform to HIPAA Privacy Rule.

4 1-Dimensional Anonymity

The results of Section 3 provide us with the right amount of generalization for each publicly known attribute in order to achieve probabilistic k-anonymity for the entire m column dataset. From any particular attribute point of view, the suggested generalization tries to create appropriate number of buckets (or partitions) in its distinct values space so that each bucket has $k' \gg k$ individuals from the universal table $\mathcal{U}$. Thus, in nutshell, there are m 1-dimensional Sweeney's k-anonymity problems, of course, each with different value of k. Before we proceed further, we will like the reader to take a note of this strong underlying connection between our notion of probabilistic k-anonymity and Sweeney's notion of k-anonymity.

Now k-anonymity for multiple columns is known to be NP-hard [26, 5, 23]. Thankfully we found that this is not the case for a single column. In the remainder of this section, we showcase various algorithms that help achieve 1-dimensional k-anonymity while retaining maximum possible data utility.

4.1 Numerical Attributes

We start out with algorithms for numerical attributes. Note that they are also applicable to attributes of type date and `Zipcode`.

Definition 6 k-Anonymous Transformation. *A k-anonymous transformation is a function, f, from $S = \{s_1, s_2, \ldots s_n\}$ to S such that $\forall s_j : |\{f^{-1}(s_j)\}| \geq k$ or $|\{f^{-1}(s_j)\}| = 0$, that is, at least k elements are mapped to each element (which has some element mapped to it) in the range.*

Example 9. *Consider $S = \{1, 12, 4, 7, 3\}$, and a function f given by $f(1) = 3, f(3) = 3, f(4) = 3, f(7) = 7$ and $f(12) = 7$. Then f is a 2-anonymous transformation.*

Dynamic Programming. Our goal is to find a k-anonymous transformation that minimizes, say, the maximum cluster size amongst all clusters [37], or the sum of distances to the cluster centers [22], or the sum over all clusters the radius of the cluster times the number of points in the cluster [6]. All these problems are known to be NP-hard for a general metric space. However, for points in a single dimension, we showcase an optimal polynomial time algorithm based on dynamic programming. The details of the algorithm can be found below.

If not already sorted, first sort the input and suppose that it is $p_1 < p_2 < \ldots < p_n$. For $1 \leq a < b \leq n$, let Cluster(a, b) be the cost to cluster elements $p_a, \ldots, p_b$.

Consider the optimal clustering of the input points. Note that each cluster in the optimal clustering contains a set of contiguous elements. Moreover, each cluster is of size at least k by the k-anonymity requirement. Since any cluster of size $\geq 2k$ can be broken into two contiguous clusters of size at least k each and that would reduce the clustering cost, the size of a cluster in the optimal clustering will be at most $2k - 1$.

The optimal clustering of the n input points is, therefore, the optimal clustering of points p_1, p_2, p_{n-i} and one single cluster of the points $(p_{n-i+1}, \ldots, p_n)$, where i is the size of the last cluster. Note that $k \leq i < 2k$ by the previous analysis. Therefore we find the optimal clustering by trying out all possible values of $i \in \{k, k+1, \ldots, 2k-1\}$. Now, the dynamic programming recursive equation is given by

ClusterCost$(1, n) = min_{k \leq i < 2k}$ Cost(ClusterCost$(1, n - i)$, Cluster$(n - i + 1, n))$.

Here Cost(A, B) is the sum for a metric like the k-median [22] or cellular [6] metric which minimizes the sum of costs over all clusters. It is the maximum function for the k-center metric [37] which minimizes the maximum of cluster sizes amongst all clusters.

ClusterCost(a, b) is initially set to ∞ if $b - a + 1 < k$. For $b - a + 1 \geq k$, ClusterCost(a, b) is initially set to the cost of clubbing all points into a single cluster, that is, Cluster(a, b).

Cost Analysis. This algorithm needs input in the sorted order. Therefore, its time complexity has two components: 1. Time taken for sorting the input, and 2. time required for the dynamic programming. For input of size n points, sorting takes $O(n \log n)$ time. The dynamic programming part requires time $O(nk)$ as evaluating ClusterCost $(1 \ldots i)$ takes $O(k)$ time for each i. Thus, overall time complexity is $O(n(k + \log n))$.

Quantiles. The algorithm from previous section requires sorting of the input. For large n, this would entail external sort. It is not very desirable in practice. In this section, we explore efficient algorithms that cluster the data in time required to make 1 or 2 sequential passes over the data and use very little extra memory.

Definition 7 Rank. *Given a set of distinct elements $S = \{s_1, s_2, \ldots, s_n\}$, the rank of an element s_i is r if s_i is the rth largest element in the set.*

For a multi-set containing duplicates, different occurrences of the same element are given consecutive ranks.

Example 10. *Among elements $S = \{1, 12, 4, 7, 3\}$, 7 has rank 4, while 3 has rank 2.*

Definition 8 Rank difference of a transformation. *Given a set $S = \{s_1, s_2, \ldots, s_n\}$ of n numbers, and a k-anonymous transformation f, let $\pi(s_i)$ represent the rank of element s_i. Then, the rank difference incurred by s_i under the transformation f is defined as $|\pi(f(s_i)) - \pi(s_i)|$. The rank difference of the transformation f is the sum of rank difference over all elements, that is, $\sum_{i=1}^{n} |\pi(f(s_i)) - \pi(s_i)|$.*

Example 11. *For set $S = \{1, 12, 4, 7, 3\}$, $\pi(1) = 1$, $\pi(12) = 5$, $\pi(4) = 3$, $\pi(7) = 4$ and $\pi(3) = 2$. For f from Example 9, $\pi(f(1)) = 2$, $\pi(f(12)) = 4$, $\pi(f(4)) = 2$, $\pi(f(7)) = 4$, and $pi(f(3)) = 2$. The rank difference of this transformation is 3.*

Definition 9 Quantile Transformation. *Suppose that $n = qk + r$, where $0 \leq r < k$. Then, the quantile transformation is a k-anonymous transformation that partitions the elements into q contiguous groups of size $(k + \lfloor r/q \rfloor)$ or $(k + \lceil r/q \rceil)$ each. All elements in a group are mapped to the median element of the group.*

Theorem 5. *The quantile transformation has the minimum rank difference among all k anonymous transformations.*

Proof. The proof is by a simple greedy argument.

Efficient Approximate Quantiles using Samples. It is possible to implement the exact quantile transformation. But finding the exact median(quantile) in p passes over the data requires $n^{1/p}$ memory [27]. Thus, to get the exact quantile transformation in 2 passes, would require $\Omega(\sqrt{n})$ memory.

For those who work with smaller memory and/or look for something easier to implement, we sketch a sampling based approach here. We maintain a uniform sample of size $s = \frac{1}{\epsilon^2} log(\frac{1}{\delta})$ using Vitter's sampling technique [38]. The rank t element in the original set is approximated by the rank st/n element in the sample, where n is the size of the original dataset over which the sample is maintained. This element has rank between $t-(\epsilon n)$ and $t+(\epsilon n)$ in the original data with probability greater than $(1-\delta)$ if the sample size s is chosen as given above [25]. For example suppose that we maintain a uniform sample of 100 elements out of a total $100,000$ elements. Then the $5,000$th element in sorted order among the $100,000$ elements can be approximated well by the 5th element in sorted order from amongst the sample of 100 elements.

4.2 Categorical Attributes

In the previous sub-section, we discussed how to create appropriate buckets or categories for numerical (ordered) attributes. But many times, there is an attribute with no intrinsic ordering among its value-set. Such an attribute is called as a *categorical attribute*.

For categorical attributes we create a layered tree graph as explained. The first layer consists of a node for each category value. The next layer groups together nodes that generalize into one general categorical value, so that they form a single node. This is set

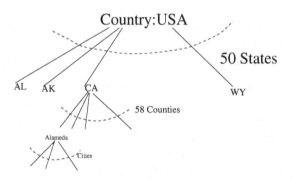

Fig. 2. A Categorical Attribute

to be the parent of the generalized values. This is repeated till there is a single category. Consider for example location information shown in Figure 2. Zipcodes are generalized to cities which are generalized to counties to state and finally to country. The top three levels of the generalization hierarchy are shown. To anonymize this dataset so that there are d distinct values, the generalization is carried till the level that there are d values. For example, to generalize location so that there are 50 different values, the state information would be retained. However to generalize it to 3000 distinct values, the county information would be retained.

5   Experiments

5.1   Quasi-identifiers

We counted the number of singletons in the Adult Database available from the UCI machine learning repository [10]. The Adult Database has 32561 rows with 15 attributes, we considered 10 of them and dropped the remaining 5. The dropped attributes are sensitive attributes (not quasi-identifiers): fnlwgt, capgain, caploss, income and the attribute edunum which is equivalent to the attribute education. In our experiments, we varied the size of the attribute set Q under consideration from 1 to the maximum of 10. The table in Figure 3 shows some of the results that we obtained.

Row	Size	A1	A2	A3	A4	A5	A6	A7	A8	A9	A10	S	F_1	D	F_2	k-Anon
		60	8	15	7	14	6	5	2	20	40					
1	1	x										2	$6.1*10^{-5}$	60	$7.4*10^{-8}$	$5*10^6$
2	2	x								x		986	0.03	1200	$1.48*10^{-6}$	$2.5*10^5$
3	3	x					x	x				65	0.002	600	$7.4*10^{-7}$	$5*10^5$
4	4	x	x	x		x						5056	0.16	$1*10^5$	$1.2*10^{-4}$	$3*10^3$
5	4	x	x			x					x	3105	0.095	$2.7*10^5$	$3.3*10^{-4}$	$1.1*10^3$
6	4	x				x				x	x	7581	0.23	$6.7*10^5$	$8.3*10^{-4}$	450
7	4		x	x		x					x	1384	0.043	$6.7*10^4$	$8.3*10^{-5}$	$4.5*10^3$
8	5	x	x	x		x					x	7659	0.235	$4*10^6$	$4.9*10^{-3}$	75
9	5	x	x		x	x	x					5215	0.16	$2.8*10^5$	$3.4*10^{-4}$	$1*10^3$
10	5	x	x			x	x			x		12870	0.40	$8*10^5$	$9.9*10^{-4}$	380
11	5	x	x			x				x	x	10402	0.32	$5.4*10^6$	$6.7*10^{-3}$	55
12	10	x	x	x	x	x	x	x	x	x	x	24802	0.76	$33*10^9$	0.99	1

Size = Number of columns that make the quasi-identifier, A1 = Age, A2 = Work class, A3 = Education, A4 = Marital status, A5 = Occupation, A6 = Relationship, A7 = Race, A8 = Sex, A9 = Hours per week, A10 = Native country, S = Number of singletons in the current table, F_1 = Fraction of singletons using the table itself = S/32561, F_2 = Fraction of singletons using Figure 1 and $n = 3*10^8$ for US population, k-Anon= Anonymity parameter for the published database = n/D.

Fig. 3. Quasi-Identifiers on the Adult Dataset

Labels **A1, A2, ..., A10** denote the 10 columns of the table. The first row gives the number of distinct values each attribute **A1, A2, ..., A10** takes. All other rows (which are labeled with row numbers from 1 to 12) of the table represent publishing the projection of the table along the columns marked 'x'. For example, the row 1 represents publishing the database projected on the Age (**A1**) column while the row 12 represents publishing all 10 columns in the database. The column **Size** gives the number of 'x' marks in each row, that is, the number of columns that constitute the quasi-identifier Q under consideration.

The column **S** is the number of rows uniquely identified by the projection of these columns, that is, the number of rows uniquely identified in the published projection. For example, for row 2, where **A1** and **A9** are the attributes of projection, **S** = 986 is returned by the following SQL statement in MS Access:

```
SELECT A1, A9 FROM T
GROUP BY A1, A9
HAVING count(*)=1
```

F_1 is the fraction of rows uniquely identified, given by $S/32561$ where S is the number of singletons while 32561 represents the total number of rows in the database table. For row 2, $F_1 = 0.03$. Some previous definitions of quasi-identifiers [40] measured a quasi-identifier as a set of columns that have a large fraction of unique rows. Thus, F_1 is used as a measure of quasiness. This does not model the external table present with the adversary. For example, by this definition, **A1** and **A9** would together be a 0.03-quasi-identifier.

D is the product of the domain sizes of the attributes marked 'x' in the row. By Multiple Domain Assumption, it is the size of the distinct values space for that combination of columns. For example, for row 3, $D = 60 * 5 * 2 = 600$.

F_2 captures the notion of quasiness as proposed in Section 2. It is given by $f(D/n)$ shown in Figure 1. Here, D is set to be equal to the value from column **D**, and $n = 3*10^8$, the size of US population. Please recall that, by Theorems 1 and 2, $f(D/n) = D/en$ for $D < n$ and $e^{-n/D}$ for $D \geq n$. For all but the last row of the table, $D < 3 * 10^8$, hence $F_2 = \frac{D}{2.7*3*10^8}$, for the last row $F_2 = e^{-3*10^8/D}$.

k-Anon is approximately the probabilistic k-anonymity obtained from the published database. Based on the result of Theorem 4, it is set to n/D, where $n = 3 * 10^8$, the size of the US population. When **D** exceeds n, it is set to 1.

Suppose we are allowed to publish a set of columns with the condition that all 0.2-quasi-identifiers are to be suppressed. If we only consider the entries of the table and look at those projections where at least 0.2 fraction of the rows are unique, then the projections indicated by rows numbered 6, 8, 10, 11 and 12 cannot be published. This is because their F_1 values exceed 0.2.

In fact, our real worry is that > 0.2 fraction of the rows should not get uniquely identified after taking an external join with the universal table $\mathcal{U}$. Then, only row 12 qualifies as a possible 0.2-quasi-identifier as only its F_2 value exceeds 0.2. Note that, from Theorems 1 and 2, there is a minuscule chance of false negatives, that is, rows $1 - 11$ are unlikely to be 0.2-quasi-identifiers.

Row 12 needs a closer look since 0.99 is only an upper bound on the expected fraction of unique rows. It may be noticed that many combinations are rare and do not occur. In our example, two attributes **A9** and **A10** are special. **A9** may be represented with only 5 distinct values since the exact hours per week of an individual may not be known and **A10** is not uniformly distributed. Such a case by case analysis of the different attributes may bring down the distinct values, **D**, and hence the fraction of distinct rows. Thus, it can help improve the estimate of quasiness, say, from a 0.99 fraction to (probably) a fraction lower than 0.2. In such a case, row 12 would be a false positive.

5.2 Anonymity Algorithms

We implemented sampling based approximate quantile algorithm (from Section 4.1) as a technique in a commercial data masking tool, MASKETEER™ [36], used at Tata Consultancy Services. Our technique only required 400 lines of code to be added to the tool, because of the extensibility features available in the tool. The tool was run on an Oracle database containing 250,000 rows of a table from a real bank, which was a customer of the tool vendor. The database table was about 1GB in size and had 261 columns. We also repeated our experiments on the public use microdata sample (PUMS) [2] provided by the U.S. Census Bureau. This dataset was given in a flat file format as input to the data masking tool. The experiments were run on a machine with 2.66GHz processor and 504 MB of RAM running Microsoft Windows XP with Service Pack 2.

Scaling with the Dataset Size

We studied how the running time of the quantile algorithm for masking a single column changes as the number of rows in the database table is varied. We measured the time required to mask various fractions of the table, the entirety of which contains 250,000 rows. The time required to mask this single numeric column with $k = 10,000$ anonymity (so that there are 25 different quantiles to which the data is approximated) increased linearly to a total of about 10 seconds for the entire column. A straight line with almost exactly identical slope and coordinates was obtained for the PUMS [2] dataset.

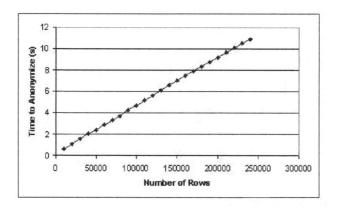

Fig. 4. Time taken for varying number of rows

Scaling with the Number of Columns Masked

We studied how the running time of the quantile algorithm for masking multiple columns varies as the number of columns to be masked is varied. For this experiment too, we used the table with 250,000 rows and 261 columns. As each column is independently anonymized, the time taken increases linearly as the number of columns being anonymized increases. Previous algorithms [23] had an exponential increase in the time taken for anonymization as the number of columns increased as the lattice created was exponential in the number of columns being anonymized.

The time taken to anonymize 10 columns of data with 250,000 rows was approximately 100 seconds. This is almost an order of magnitude improvement over the previous algorithm [23]. The results on the PUMS dataset were similar.

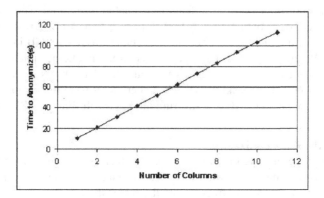

Fig. 5. Time taken for varying number of columns

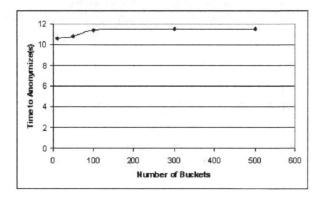

Fig. 6. Time taken for varying number of buckets

Scaling with the Anonymity Parameter

The implemented algorithm does a binary scan over all buckets to find the bucket closest to each data item. The time required to anonymize a data value, therefore, logarithmically increases as the number of buckets increases (or the value k of anonymity parameter decreases). If b is the number of buckets and n the number of rows, then the time to anonymize is $nlog(b)$. The time taken to read n rows from disk is nC where C is a large constant[3]. The total time taken is, therefore, $n(C + \log b)$ where $C \gg \log(b)$. This explains the shape of the curve in Figure 6. Here $nC \approx 10$ seconds and the $log(b)$ term explains the slight increase from 0 to 500 buckets.

Tradeoff between Privacy and Utility

We use the term error for information loss introduced by the anonymization process. We studied how the error introduced in a column as a result of k-anonymization varies with

---

[3] Transfer rate of 50MBps is not uncommon for harddisks today. So if each record were to be of size 1KB, then $C = 1/50000$.

the anonymity parameter k. Let x_i be the original value of the ith row. Let x_i' be its value after k-anonymization. Let us use $(x_i' - x_i)^2$ as the error introduced for row i as a result of k-anonymization. The total error introduced over n rows is $Error = \sum_{i=1}^{n}(x_i' - x_i)^2$. Let $\bar{x} = \frac{\sum_{i=1}^{n} x_i}{n}$. If all x_i' are constrained to be identical (corresponding to anonymity with a single bucket), then $\bar{x}$ gives the minimum error according to the above metric, i.e. it gives $MinError = Min_x \sum_{i=1}^{n}(x - x_i)^2 = \sum_{i=1}^{n}(\bar{x} - x_i)^2$. We, therefore, normalize the error as Error/MinError.

The curve is plotted in Figure 7 where the normalized error is plotted on the y-axis while the number of buckets, $b = \lceil \frac{n}{k} \rceil$, is plotted on the x-axis. An almost identical curve was obtained for the PUMS dataset. The curve very closely follows the curve $\frac{1}{b^2}$. This could be proven analytically.

Thus, for given n and k, we find that the identity disclosure risk is $< 1/k$ (for "join" class of attacks) and the error introduced in data is $\propto k^2/n^2$. We may, therefore, boldly quantify the privacy provided by k-anonymization as $p = 1 - 1/k$ and the utility retained as $u = 1 - k^2/n^2$ implying the following privacy-utility trade-off equation.

$$(1 - p)^2(1 - u) = 1/n^2 \text{ (a constant)}.$$

Note that, the fact that we used sum square errors, instead of sums of absolute values of errors explains the square term above.

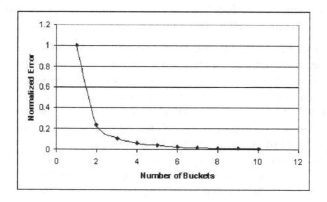

Fig. 7. Tradeoff between privacy and utility

6   Related Work

One of the earliest definitions of quasi-identifier can be found in Dalenius [16]. [35,34] and [23] use a similar definition.

Samarati and Sweeney formulated the k-anonymity framework and suggested mechanisms for k-anonymization using the ideas of generalization and suppression [29, 35,34]. Subsequent work has shown some NP-hardness results [26,4,6] and that has inspired many interesting heuristics and approximation algorithms [21, 39, 26, 9, 4, 23, 24, 6]. All of this work assumes that quasi-identifier attribute sets are known based on specific knowledge domain.

The basic theme of k-anonymity model is to *hide* an individual in a crowd of size k or more. A similar intuition is pursued by Chawla et al in [13] who, in fact, manage to convert it into a precise mathematical statement. They not only give a definition of privacy and its compromise for statistical databases, but also provide a method for describing and comparing the privacy offered by specific sanitization techniques. They also give a formal definition of an *isolating* adversary whose goal is to single out someone from the crowd with the help of some *auxiliary* information z. This work is further extended in [14] where Chawla et al study privacy-preserving histogram transformations that provide substantial utility.

There is a wide consensus that privacy is a corporate responsibility [20]. In order to help and ensure corporations fulfil this responsibility, governments all over the world have passed multiple privacy acts and laws, for example, Gramm-Leach-Bliley (GLB)Act [18], Sarbanes-Oxley (SOX) Act [30], Health Insurance Portability and Accountability Act (HIPAA) [19] Privacy Rule are some such well known U.S. privacy acts. In fact, HIPAA recommends the following *safe-harbor* method of de-identification in which it provides clear guidelines for sanitizing quasi-identifiers including date types, Zipcode, etc. For 20, 000 anonymity, HIPAA advises to retain essentially only the State information in Zipcode and year information in Date of Birth which is quite inline with what we concluded in Examples 6, 7 and 8 based on our analysis. The de-identification excerpt from the HIPAA privacy rule is provided below.

"The following identifiers of the individual or of relatives, employers, or household members of the individual must be removed to achieve the "safe harbor" method of de-identification: (A) Names; (B) All geographic subdivisions smaller than a State, including street address, city, county, precinct, zip code, and their equivalent geocodes, except for the initial three digits of a zip code if, according to the current publicly available data from the Bureau of Census (1) the geographic units formed by combining all zip codes with the same three initial digits contains more than 20,000 people; and (2) the initial three digits of a zip code for all such geographic units containing 20,000 or fewer people is changed to 000; (C) All elements of dates (except year) for dates directly related to the individual, including birth date, admission date, discharge date, date of death; and all ages over 89 and all elements of dates (including year) indicative of such age, except that such ages and elements may be aggregated into a single category of age 90 or older; (D) Telephone numbers; (E) Fax numbers; (F) Electronic mail addresses: (G) Social security numbers; (H) Medical record numbers; (I) Health plan beneficiary numbers; (J) Account numbers; (K) Certificate/license numbers; (L) Vehicle identifiers and serial numbers, including license plate numbers; (M) Device identifiers and serial numbers; (N) Web Universal Resource Locators (URLs); (O) Internet Protocol (IP) address numbers; (P) Biometric identifiers, including finger and voice prints; (Q) Full face photographic images and any comparable images; and (R) any other unique identifying number, characteristic, or code, except as permitted for re-identification purposes provided certain conditions are met. In addition to the removal of the above-stated identifiers, the covered entity may not have actual

knowledge that the remaining information could be used alone or in combination with any other information to identify an individual who is subject of the information. 45 C.F.R. §164.514(b). "

7 Conclusions

In this paper, we provided the first formalism and a practical technique to identify a quasi-identifier. Along the way we discovered an interesting connection between whether a set of columns forms a quasi-identifier and the number of distinct values assumed by the combination of the columns.

Then we defined a new notion of anonymity called as probabilistic anonymity where in we insist that each row of the anonymized dataset should match with at least k or more rows of the universal table $\mathcal{U}$ along a quasi-identifier. We observed that this new notion of anonymity is similar to the existent k-anonymity notion in terms of privacy guarantees and is sufficiently strong for many real life scenarios involving oblivious adversaries. Building on our earlier work, we found an interesting connection between the number of distinct values taken by a combination of columns and the anonymity it can offer. This allowed us to find an ideal amount of generalization or suppression to apply to different columns in order to achieve probabilistic anonymity. We worked through many examples and showed that our analysis can be used to make a published database conform to privacy rules like HIPAA.

In order to achieve the probabilistic anonymity, we observed that one needs to solve multiple 1-dimensional k-anonymity problems. We proposed many efficient and scalable algorithms for achieving 1-dimensional anonymity. Our algorithms are optimal in a sense that they minimally distort data and retain much of its utility.

Acknowledgements

We thank Ravishankar Rajamony and Ellora Praharaj for Section 5.1 and the entire MASKETEER™ team, namely, Prasenjit Das, Anshula Dhar, Rahul Ghodeswar, Sumit Johri, Snighda Nayak, Jayant Patil, Nikhil Patwardhan, Ashim Roy, Bharath S, Rupali Sagade, Aparna Sarnobat and Sharada Sundaram for building the MASKETEER™ tool on which our algorithms were implemented. The second author would also like to thank the people at TRDDC for hosting him for a pleasant summer. He would also like to thank Jeff Ullman and Rajeev Motwani for ideas that helped shape the paper.

References

1. Accuracy of the US census data, U.S. Census Bereau,
   http://www.census.gov/acs/www/UseData/Accuracy/Accuracy1.htm
2. Public use microdata sample (PUMS), U.S. Census Bureau,
   http://www.census.gov/acs/www/Products/PUMS/
3. Aggarwal, C.C.: On k-anonymity and the curse of dimensionality. In: Proceedings of the 2005 International Conference on Very Large Data Bases, pp. 901–909 (2005)

4. Aggarwal, G., Feder, T., Kenthapadi, K., Motwani, R., Panigrahy, R., Thomas, D., Zhu, A.: Anonymizing tables. In: Proceedings of the International Conference on Database Theory, pp. 246–258 (2005)

5. Aggarwal, G., Feder, T., Kenthapadi, K., Motwani, R., Panigrahy, R., Thomas, D., Zhu, A.: Approximation algorithms for k-Anonymity. Journal of Privacy Technology, 20051120001 (2005); Earlier version appeared in Proc. of the Intl. Conf. on Database Theory (ICDT 2005)

6. Aggarwal, G., Feder, T., Kenthapadi, K., Panigrahy, R., Thomas, D., Zhu, A.: Clustering for privacy. In: Proceedings of the ACM Symposium on Principles of Database Systems (2006)

7. Agrawal, R., Srikant, R.: Fast Algorithms for Mining Association Rules. In: Proceedings of the International Conference on Very Large Data Bases, Santiago, Chile, pp. 487–499 (September 1994)

8. Baum, K.: First estimates from the national crime victimization survey: Identity theft, 2004. In: Bureau of Justice Statistics Bulletin (April 2006),
http://www.ojp.usdoj.gov/bjs/pub/pdf/it04.pdf

9. Bayardo, R.J., Agrawal, R.: Data privacy through optimal k-anonymization. In: Proceedings of the International Conference on Data Engineering, pp. 217–228 (2005)

10. Blake, C., Merz, C.: UCI repository of machine learning databases (1998),
http://www.ics.uci.edu/~mlearn/MLRepository.html

11. Brown, M.: Identity theft victim stories: Verbal testimony by michelle brown. In: Privacy Rights Clearing House (July 2000),
http://www.privacyrights.org/cases/victim9.htm

12. Chaudhuri, S., Ganjam, K., Ganti, V., Motwani, R.: Robust and efficient fuzzy match for online data cleaning. In: Proceedings of the ACM SIGMOD International Conference on Management of Data (2003)

13. Chawla, S., Dwork, C., McSherry, F., Smith, A., Wee, H.: Toward privacy in public databases. In: 2nd Theory of Cryptography Conference (TCC), pp. 363–385 (2005)

14. Chawla, S., Dwork, C., McSherry, F., Talwar, K.: On the utility of privacy-preserving histograms. In: 21st Conference on Uncertainty in Artificial Intelligence (UAI) (2005)

15. Chernoff, H.: Asymptotic efficiency for tests based on the sums of observations. Annals of Mathematical Statistics 23, 493–507 (1952)

16. Dalenius, T.: Finding a needle in a haystack or identifying anonymous census records. Journal of Official Statistics (2), 329–336 (1986)

17. Gibbons, P.B.: Distinct sampling for highly-accurate answers to distinct values queries and event reports. In: Proceedings of the International Conference on Very Large Data Bases, pp. 541–550 (2001)

18. GLB. Gramm-Leach-Bliley Act,
http://www.ftc.gov/privacy/privacyinitiatives/glbact.html

19. HIPAA. Health Information Portability and Accountability Act,
http://www.hhs.gov/ocr/hipaa/

20. IBM. Privacy is good for business,
http://www-306.ibm.com/innovation/us/customerloyalty/
harriet_pearson_interview.shtml

21. Iyengar, V.: Transforming data to satisfy privacy constraints. In: 8th ACM SIGKDD International Conference on Knowledge Discovery in Databases and Data Mining, pp. 279–288 (2002)

22. Jain, K., Vazirani, V.: Primal-dual approximation algorithms for metric facility location and k-median problems. In: Proceedings of the Annual IEEE Symposium on Foundations of Computer Science, pp. 2–13 (1999)

23. Lefevre, K., Dewitt, D.J., Ramakrishnan, R.: Incognito: Efficient full domain k-anonymity. In: Proceedings of the ACM SIGMOD International Conference on Management of Data, pp. 49–60 (2005)

24. Machanavajjhala, A., Kifer, D., Gehrke, J., Venkitasubramaniam, M.: l-diversity: Privacy beyond k-anonymity. In: Proceedings of the International Conference on Data Engineering, p. 24 (2006)

25. Manku, G.S., Rajagopalan, S., Lindsay, B.G.: Random sampling techniques for space efficient online computation of order statistics of large datasets. In: Proceedings of the ACM SIGMOD International Conference on Management of Data, pp. 251–262 (1999)

26. Meyerson, A., Williams, R.: On the complexity of optimal k-anonymity. In: Proceedings of the ACM Symposium on Principles of Database Systems, pp. 223–228 (June 2004)

27. Munro, I., Paterson, M.: Selection and sorting with limited storage. In: Proceedings of the Annual IEEE Symposium on Foundations of Computer Science, pp. 253–258 (1978)

28. Rudin, W.: Real and Complex Analysis. McGraw-Hill, New York (1987)

29. Samarati, P., Sweeney, L.: Generalizing data to provide anonymity when disclosing information (abstract). In: Proceedings of the ACM Symposium on Principles of Database Systems, p. 188 (1998)

30. SOX. Sarbanes-Oxley Act, http://www.sec.gov/about/laws/soa2002.pdf

31. Sweeney, L.: Guaranteeing anonymity when sharing medical data, the datafly system. In: Proceedings of the Journal of the American Medical Informatics Association Annual Fall Symposium, pp. 51–55 (1997)

32. Sweeney, L.: Three computational systems for disclosing medical data in the year 1999. In: Proceedings of MEDINFO, pp. 1124–1129 (1998)

33. Sweeney, L.: Uniqueness of simple demographics in the U.S. population. In: LIDAP-WP4. Carnegie Mellon University, Laboratory for International Data Privacy, Pittsburgh, PA (2000)

34. Sweeney, L.: Achieving k-anonymity privacy protection using generalization and suppresion. International Journal on Uncertainty, Fuzziness and Knowledge-based Systems 10(5), 571–588 (2002)

35. Sweeney, L.: k-Anonymity: A model for preserving privacy. International Journal on Uncertainty, Fuzziness and Knowledge-based Systems 10(5), 557–570 (2002)

36. TRDDC. Masketeer: A tool for preserving privacy, Pune (2005)

37. Vazirani, V.: Approximation Algorithms. Springer, Heidelberg (2004)

38. Vitter, J.: Random sampling with a reservoir. In: ACM Transaction on Mathematical Software, pp. 37–57 (1985)

39. Winkler, W.: Using simulated annealing for k-anonymity. In: Research Report 2002-07, US Census Bureau Statistical Research Division (November 2002)

40. Xu, Y., Motwani, R.: Random sampling based algorithms for efficient semi-key discovery (2006),
http://theory.stanford.edu/~xuying/papers/minkey_vldb.pdf

Website Privacy Preservation for Query Log Publishing

Barbara Poblete[1], Myra Spiliopoulou[2], and Ricardo Baeza-Yates[1,3]

[1] Web Research Group, University Pompeu Fabra, Barcelona, Spain
[2] Otto-von-Guericke-University Magdeburg, Germany
[3] Yahoo! Research, Barcelona, Spain
barbara.poblete@upf.edu,
myra@iti.cs.uni-magdeburg.de, ricardo@baeza.cl

Abstract. In this paper we study privacy preservation for the publication of search engine query logs. We introduce a new privacy concern, *website privacy* as a special case of *business privacy*. We define the possible adversaries who could be interested in disclosing website information and the vulnerabilities in the query log, which they could exploit. We elaborate on anonymization techniques to protect website information, discuss different types of attacks that an adversary could use and propose an anonymization strategy for one of these attacks. We then present a graph-based heuristic to validate the effectiveness of our anonymization method and perform an experimental evaluation of this approach. Our experimental results show that the query log can be appropriately anonymized against the specific attack, while retaining a significant volume of useful data.

1 Introduction

Query logs are very rich sources of information, from which the scientific community can benefit immensely. These logs allow among other things the discovery of interesting behavior patterns and rules. These can be used in turn for sophisticated user models, for improvements in ranking, for spam detection and other useful applications. However, the publication of query logs raises serious and well-justified privacy concerns: It has been demonstrated that naively anonymized query logs pose too great a risk in disclosing private information.

The awareness towards privacy threats has increased by the publication of the American Online (AOL) query log in 2006 [1]. This dataset, which contained 20 million Web queries from $650,000$ AOL users, was subjected to a rather rudimentary anonymization before being published. After its release, it turned out that the users appearing in the log had issued queries that disclosed their identity either directly or in combination with other searches [2]. Some users even had their identities published along with their queries [3]. This increased the awareness to the fact that query logs can be manipulated in order to reveal private information if published without proper anonymization.

Privacy preservation in query logs is a very current scientific challenge. Some solutions have been proposed recently [4,5]. Similarly to the general research

F. Bonchi et al. (Eds.): PinKDD 2007, LNCS 4890, pp. 80–96, 2008.

advances in privacy preserving data mining, they refer to the *privacy of persons*. Little attention has been paid to another type of privacy concern, which we consider of no less importance: *website privacy* or, more general, *business privacy*.

In this work we argue that important and confidential information *about websites and their owners* can be discovered from query logs and that naive forms of URL anonymization, as in [2], are not sufficient to prevent adversarial attacks. Examples of information that can be revealed from query logs include accesses to the site's documents, queries posed to reach these documents and query keywords that reflect the market placement of the business that owns the site. Such pieces of information *are* confidential, because websites serve as channels for advertisement, communication with potential customers and often sales to them. Hence, the traffic recorded in them delivers a picture of customer-company interaction, possibly for the whole product portfolio. A thorough analysis of this traffic with a data mining method may then deliver information like insights on the effectiveness of advertising campaigns, popular and less popular products, number of successful and failed sale transactions etc.

One may argue that a site's traffic is only recorded at the site's server and therefore not public. However, the traffic delivered to a website by major search engines accounts for an important part of the site's overall traffic. If this part is undisclosed, it will be a very close approximation to the complete access log of the website.

The protection of such confidential information is different from conventional privacy preservation. One reason for this difference is that an adversary can reveal confidential website information by aggregating a published query log with other legally owned private data. In particular, consider an adversary which is a company interested in disclosing information about its competitors. This adversary could use its own background knowledge and the data of its own site in combination to the published query log data, to infer the competitor's private data. This includes but is not limited to popular queries that reach *both* the adversary's site and that of the competitor. As shown in Fig. 1, the log of the adversary can then be used to de-anonymize a part of the published query log. Depending on the amount and quality of the information revealed, *industrial espionage* or malicious intent could be argued by the affected parties against the company that published the query log.

Although query log anonymization does not look promising in the near future, especially from the user privacy perspective, we believe that reasonable measures can be taken to preserve website privacy. By discussing some of the existing threats and ways to prevent them, we can set a precedent for data mining applications on logs, and future query log publishing, so that the resulting information is inspected to prevent privacy leaks. Although we focus on website privacy, we believe that our approach also contributes to user privacy, because much of the sensitive information about users comes from assessing the pages they have visited.

The contributions of our work are as follows: (1) We introduce a new privacy issue for query logs, *website privacy*. (2) We describe attacks that disclose confidential information from query logs, and ways to prevent them. (3) We propose a heuristic graph-based method that removes those parts of the log that may lead to information disclosure and we validate it with experiments over real data.

In the next section, we discuss related work on privacy preserving data mining and on query log anonymization for the protection of *user privacy*. In section 3, we introduce the problem of website privacy preservation through anonymization and describe the types of adversaries that might attack a published query log. Section 4 describes attacks and counter-measures. In section 5, we implement a counter-measure with an heuristic that eliminates the vulnerable parts of the query log. We validate this heuristic experimentally and report our results in section 6. The last section concludes the study with a summary and a short discussion of open issues.

2   Related Work

The rapid development of advanced techniques for data collection and propagation, along with the fast growth of the Web, have increased the awareness to the use of private information. This has lead to a new field of research in the context of analyzing private or confidential information – the domain of *privacy preserving data mining* [6].

Privacy preserving data mining aims at analyzing databases and data mining algorithms, identifying potential privacy violations and devising methods that prevent privacy breaches. Preventive measures involve the hiding or modification of sensitive raw data like names, credit card numbers and addresses, and the exclusion of any further type of sensitive knowledge that can be mined from the database. It is important to note that many privacy preserving algorithms are based on heuristics. This is because of the premise that selective data modification or sanitization is an NP-hard problem.

The evaluation of privacy preserving algorithms [6] is usually centered on the following features: the *performance* of the proposed algorithm, the *data utility* after the application of the technique, the *level of uncertainty* which the sensitive information can be predicted, and the *resistance* to different data mining techniques.

Some research on privacy preservation in databases deals with privacy preserving data publishing that guarantees utility for data mining [7,8]. There are studies on preventing adversarial data mining in relational databases, when data fields are correlated [9]. Samarati and Sweeney proposed *k-anonymity*, in which data is released in such a way that each query result (and each attempt for data disclosure) returns at least k entities [10]. The principle of k-anonymity is quite effective but it cannot be directly applied to data that expands across multiple databases, as is the case of website privacy preservation.

In the context of Web mining, one of the prominent areas for privacy preservation is the protection of user privacy in query logs of search engines. Among the advances in privacy preserving Web mining, most relevant to our work are the studies of Kumar et al [4] and of Adar [5]. Kumar et al propose token-based hashing for query log anonymization [4]; The queries are tokenized and a secure hash function is applied to each token. However, the authors show how statistical techniques can be used to disclose private information despite the anonymization; they also show that there is no satisfying framework to provide privacy in query logs [4].

In [5], Adar explains many aspects of the AOL query log problem, and shows that traditional privacy preservation techniques cannot be applied in a straight-forward way to protect privacy in a search log. Further, Adar argues that k-anonymity is too costly for rapidly changing datasets like query logs. Then, Adar proposes two user anonymization methods for query logs, which attempt to balance log utility for research and privacy [5].

To the best of our knowledge ours is the first paper to address the issue of privacy preservation for websites or businesses in query logs. As we will explain in the next section, website privacy preservation is a different problem than user privacy preservation. An anonymization method that preserves user privacy would not necessarily guarantee website privacy.

3 The Website Anonymization Problem

For our analysis, we assume that a query log contains at least the same fields as the one published by AOL [1]. Therefore we define our default query log format as:

$$\{AnonID,\ Query,\ QueryTime,\ ItemRank,\ ClickURL\}$$

In this signature, *AnonID* refers to an anonymized user ID, *Query* is the search string, *QueryTime* is the time at which the query was issued, *ItemRank* is the rank of the document clicked as a result of the query, and *ClickURL* is the *truncated* URL of the document. In the AOL log, the URLs of documents were truncated up to the website name ((e.g. `www.example.org/somepage.html` became `www.example.org`). For our analysis, we also consider the hostname as a central concept and define a *website* as a set of pages under the same hostname. We use this signature as a reference basis but point out that it its vulnerability with respect to user IDs has already been shown [5].

3.1 Challenges for Query Log Anonymization

Anonymizing query logs for data mining is very challenging for several reasons. First, the attributes of the query log are not independent. An adversary may use these dependencies to deduce the value of an anonymized field. For example, queries in search engines are known to exhibit a remarkable frequency distribution: Kumar et al exploited this property to decrypt anonymized queries by studying the frequency and co-occurrence of terms in a non-anonymized reference log [4]. Moreover, query logs have *sequential* records: Rearranging or shuffling them for anonymization purposes would blur or eliminate important temporal and order-dependent information, such as user sessions.

Despite these observations, we should keep in mind that data mining focuses mostly on extracting knowledge in pattern form and does not always require exact values for each attribute: Such values can be replaced by an anonymized value that preserves their distribution. However, for Web query mining, it is difficult to determine which attributes should be anonymized or hidden: *all* attributes in the log are of potential use – depending on the purpose of the analysis. Thus, the minimization of the private information that could be disclosed by an adversary while maintaining enough information for data mining becomes a complex optimization problem. We address this problem in a conservative heuristic way: We describe possible anonymizations, show attacks that can be used to disclose private data despite these anonymizations and then we increase the level of information hiding to prevent information disclosure.

3.2   Types of Adversaries for Website Privacy Preservation

It is important to recapitulate the scope of our work here: we describe and focus *only* on the problem of *website privacy preservation* or prevention of *website exposure* (singling-out a website), when publishing or sharing search engine query logs. This means that our goal is to prevent an adversary from obtaining confidential information about traffic to websites, which have been recorded in a search engine's query logs. For this objective of website privacy preservation, we identify two types of adversaries:

1. **General Adversary:** This type of adversary is "just" trying to discover useful information about as many websites as possible, without any particular site in mind. This type of adversary might show up as a search engine optimization company or other institution that performs market studies.
2. **Adversarial Competitor:** This type of adversary is a website or company that tries to disclose information about its competitors using the query log. In many cases, this adversary has already some information about the market share, portfolio and activities of the competitors, and can impute this background knowledge upon the anonymized log to de-anonymize it. One of the most important pieces of data that this adversary can exploit is its own query log, which can be used to recreate pieces of the anonymized query log.

3.3   Data Sources in the Website Privacy Preservation Problem

The query log is not the sole data source available to an adversary. We enumerate here the data sources which might be combined to assess additional information or disclose private data. The sources are depicted in Fig. 1. Some of them may be publicly available, while others may be private. They include:

1. An *anonymized search engine query log*: This is the published log.
2. Actual *search results for queries* from a search engine: This information is available to each user issuing a query. The results may come from the same search engine as the log or from a different one, which has similar document coverage.

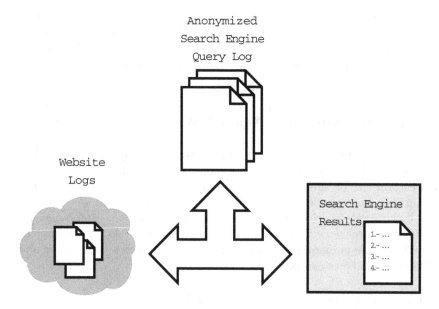

Fig. 1. Different data sources involved in query log privacy preservation

3. The *access log of a given site*: This is private to the owner of the site. The log contains clicks from external search engines. It is quite straightforward to reconstruct the origin of a click from the click's *referer*[1] in a conventional access log. An adversarial competitor will have such a log for its own website.

Example 1. To illustrate the challenges of combining sites for data mining, let us consider an on-line computer hardware store "Site A". Site A knows that the most popular queries used to reach it are: *refurbished computer, cheap notebook, laptop, memory* and *desktop computer*. Site A has access to a published search engine query log (such as the AOL log), and it wants to discover as much information as possible about its competitors. It first assumes that its competitors are reached via the same queries, which form an initial list of keywords L_0. This initial list can be expanded by searching for the URLs of Site A's most important competitor "Site B" in the query log. This allows Site A to obtain additional keywords from Site B, such as *electronics store, portable computers* and *computer deals*. The result is an enhanced list of keyword terms L_1.

Next, Site A issues the queries in L_1 online to a search engine and discovers which *other sites* are returned in each result-set. Site A discovers that its competitors (next to Site B) are Site C and Site D. Having found out these additional competitor sites, Site A can now extract all traffic data for each of them from the published query log. Among other things, Site A can discover which site has more visitors and more hits from the search engine, which queries reach other sites but do not reach Site A etc.

---

[1] This is a misspelling of "referrer". It is the official term used in HTTP specifications.

With this information, Site A can now advertise exactly on the most popular keywords used by people reaching sites B, C and D, or it may focus on only one of the competitors in a similar way. Most importantly, Site A can make business decisions based on information that was not available before the disclosure of the log.   □

4   Attacks and Measures Against Website Disclosure

In this section we discuss an incremental approach to anonymize a query log. In each step, we show how an adversary could go about to discover information about a certain site or site-related information (i.e. who are its competitors). In our analysis, we consider three main types of attacks, which we define as *attacks on vulnerable queries* (defined in subsection 4.2 below), *attacks using a website log* and *attacks with user information*. Although we do not attempt to identify *all* possible vulnerabilities, we show that several weak points exist and that different techniques can be used to prevent them.

4.1   Structure of the Anonymized Query Log

As explained before, the anonymization used in the AOL log was not sufficient for user and website privacy preservation. Although the clicked URLs were truncated at the site level, the log still provided the original *Query* and *ItemRank* information. This is enough to do a look-up on the AOL search engine and discover most of the actual URLs for each *ClickURL* field. As a result, all site information available on the original query log was disclosed.

A simplistic approach to prevent this kind of information disclosure would be to hide the *Rank* attribute in the log and at the same time do a simple anonymization on the *ClickURL*. The idea here is to replace this attribute with a unique identifier. This can be done using three methods:

1. Assign a unique ID to each URL: By doing this, the information that some URLs belong to the same domain is lost.
2. Assign a unique ID to each URL *and* distinguish among URLs that belong to the same site or domain: This can be done e.g. by using the same suffix ID for all URLs of the same site. This delivers more information than the method above, so that one can e.g. calculate the number of clicked pages in a site and the occasions, under which many different pages of the same site were clicked.
3. Assign the same ID to all URLs in a website: This still allows an analysis of the click distribution and other useful statistics for rank experiments, but it does not allow documents inside a website to be set apart.

We opt for the second method for URL anonymization. It allows us to preserve most information about the website and the accesses in its pages.

4.2   Attacks on Vulnerable Queries

We first consider the scenario of an adversary who only has access to public data sources, such as the anonymized query log and a public search engine. This corresponds to the typical *general adversary*. In this scenario, one can obtain information about any website by exploiting certain types of *"vulnerable"* queries. We use this term for queries whose results disclose directly the identity of the website; the adversary does not need access to additional information sources.

A first type of vulnerable queries are those that contain the target URL as keyword. This is a subcategory of so-called *"navigational queries"* [11]: These are queries for which the users know exactly the page they want to reach and use the search engine to obtain the URL, using it like a bookmark on a browser. Navigational queries become *"vulnerable"* from a privacy preservation point of view when they include only the terms that later appear in the root of the selected URL (i.e. the website root). For example, a query with the term "amazon" becomes vulnerable, if the user selects http://www.amazon.com among the results. Hence, the adversary can discover the actual website, even if the log is anonymized. To prevent this, the query log should be checked for queries that contain keywords matching the URL root string. Such queries should be removed or the keywords should be hidden.

Another type of *vulnerable queries* are those that return fewer than k results and thus prevent an anonymization satisfying k-anonymity. The value of k is application-specific. Once this value is set, all queries returning less than k results must be removed.

The last and more complex type of attack using *vulnerable queries* involves pairs of queries that have non-empty intersections among the *clicked* results. For this scenario, we first assume an *adversarial competitor* who tries to find information about specific sites; later, we generalize to both types of adversary. The attack of the adversarial competitor can go as follows:

1. The adversary defines a set of queries Q_1, which are known to return URLs of the competitor websites at highly ranked positions.
2. The adversary performs a look-up of the occurrences of the queries in Q_1 in the anonymized query log L and obtains a set of *AnonID* values, i.e. anonymized URLs. These IDs constitute the set of "candidate competitors" CC, i.e. initially all sites in L that are in the result of Q_1 are possible competitors.

    The task at hand is to map as many *AnonIDs* in CC as possible to the corresponding URLs. Once the URLs are known, all relevant information for them (and ultimately for the whole site containing them) can be extracted from L.
3. For each *AnonID* $u \in CC$, the adversary collects *all* unique queries that have u as a clicked URL in L. We call this set Q.
4. For each $q_i \in Q$ the adversary collects the *anonymized result-set* R_{A_i}.

5. For each pair of queries $\{q_i, q_j\} \subseteq Q$, such that $|R_{Ai} \cap R_{Aj}| \geq 1$, the adversary issues both of them live to the search engine and recovers their *real result-sets* R_i and R_j.

   If $|R_{Ai} \cap R_{Aj}| = |R_i \cap R_j| \geq 1$, then it is known that the URLs in $|R_{Ai} \cap R_{Aj}|$ have been *approximately* mapped to real URLs. The match becomes *exact* if $|R_{Ai} \cap R_{Aj}| = 1$, or if all but one URLs in $|R_{Ai} \cap R_{Aj}|$ have already been disclosed using the same methodology.

This attack can be extended for a *general adversary*: The complete log must be scanned to build the intersection of clicked URLs in two result-sets. The queries contributing to this intersection must be added to Q. Then, steps 4 and 5 are applied as above.

The process above does not guarantee that the adversary will disclose all URLs of potential interest, but will nonetheless disclose *all* of the URLs from each affected website. To alleviate this vulnerability, we propose to remove one of any two queries that share at least one clicked result. In Section 5 we present a heuristic that modifies the query log to this extent.

4.3   Attacks Using a Website Log

In the previous scenario, we assumed that the adversary had only access to the anonymized log and to the results of a public search engine. Now we turn to a scenario, where the adversary owns a website and can therefore use its access log.

The access log of a website registers all requests towards the site, including the requested URL, the time of the request, agent and IP address of the user. In the case of combined logs, the URL where the request was performed is also recorded (the referrer URL). If the referee is a search engine, the access log of the website will also contain the keywords and the URL of the search engine.

Hence, if the adversary has access to a website log of the same period as the anonymized query log, then the adversary can combine the private website log and the public query log to disclose the anonymized information of the latter. For example, the adversary can find the *AnonId* assigned to the own website (pages). In many cases, this information is not adequate for a privacy breach. However, if many sites collude and share their logs, then these logs can be combined to undisclose URLs and launch the attack described in the previous subsection.

To avoid this scenario, one more constraint should be placed in the anonymization process of the query log: The results displayed by the search engine for any given query *must contain URLs of at least k different sites*, so that k-anonymity can be pursued. Since this scenario requires collusion of multiple adversaries, we do not discuss it further but concentrate on the simpler scenario that involve only one adversary.

4.4   Attacks with User Information

We finally consider the scenario in which the adversary can disclose the identity of a (single) user in the query log. Here, we assume that identity disclosure

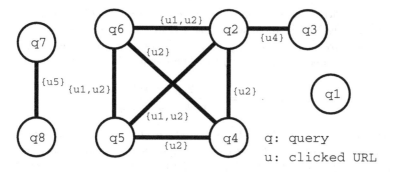

Fig. 2. Graph representation of a tiny example query log

also implies disclosure of the results clicked by the compromised user. Then, the adversary can trace the user and the clicked URLs in the anonymized log and map the *AnonID* values to complete URLs. Such an attack can be manifested by having a particular user or agent submit queries to the search engine regularly and then trace back the queries and their results in a periodically released anonymized search log. If the queries involve pages of the adversary's competitors, their results can be exploited to perform the attack described in subsection 4.2.

This scenario can be suppressed by preventing the identification of individual users in the query log. As pointed out in Section 2, this is a separate problem, for which solutions start emerging.

5   Graph-Based Method

The attack presented in subsection 4.2 exploits the occurrence of the same URL(s) among the clicked results of different queries. We have designed a graph-based method to analyze the vulnerability of query logs to this attack. At the same time we use this method to see how the number of intersections between clicked result-sets of queries can be reduced.

This graph representation of the query log consists of modeling queries as nodes. Two nodes are connected with an undirected edge *if they share at least one URL between their clicked result-sets*. This means that two queries are connected by an edge if there exist one or more URLs clicked from both queries, as shown in Fig. 2.

Our graph representation also takes into account the fact that not all nodes are of equal importance for data mining applications. To represent the "value" of each node in the log, we assign weights to the queries. For example, the weight can reflect the frequency of a particular query in the log or the number of clicked documents for that query.

Using this graph approach we show that the solution to the attack described in subsection 4.2 is a *well-defined optimization problem*, namely that of disconnecting the graph by removing nodes while preserving the maximum weighted graph. This corresponds to finding the *maximum (weighted) independent set.* This problem is NP-Hard, so we define a heuristic approach. For this, we first define a measure of *graph density,* which reflects *how likely it is to find an edge among any two nodes in the query log.* The formula for this measure is:

$$Density = \frac{2(\# \ edges)}{\# \ nodes \ (\# \ nodes\text{-}1)}$$

Then, the goal of our heuristic is to reduce *Density* for the query log graph to zero. The zero value of *Density* means that there are no intersections among result-sets in the log. The challenge for the heuristic is to disconnect the graph while attempting to preserve the nodes with maximum weight.

If the number of edges per node in the graph follow a power law, then this would indicate that the number of edges can be rapidly reduced by node removal: The graph would become disconnected very fast by only removing a few high-degree nodes in the graph [12]. With this in mind, we define the following greedy heuristic to disconnect the graph:

1. Sort nodes by their *degree.*
2. Remove the node with the *highest degree.*
3. Recalculate the degree of all nodes that were adjacent to the node that was removed.
4. Compute the value of *Density.*
5. If the value of *Density* is not equal to zero, then go to Step 1, otherwise finish.

This heuristic can be extended to incorporate the weight of each node. This is done by replacing the criterion used in Step 1 to sort the nodes, and use $\frac{degree}{weight}$ instead of *degree.*

Once the graph is disconnected, certain characteristics can be analyzed retrospectively, such as the speed at which the value of *Density* decreased and the number of queries and of clicked documents that had to be removed to completely disconnect the graph.

6   Experimental Results and Discussion

6.1   The Dataset for the Evaluation

For our evaluation, we used a query log from the Yahoo! search engine. For privacy reasons these logs are carefully controlled and cannot be released for general study. Even for this analysis, we do not deal with the raw query log, but only with its graph representation. The graph representation is an application of the graph models developed in [13] and can be computed rapidly. The computation took approximately 2 hours on a dual core AMD Opteron^{TM} Processor 270 with 6.7 gigabytes of RAM; it is noted though that the processing always used less than 4 gigabytes of memory and employs only one CPU. The resulting graph

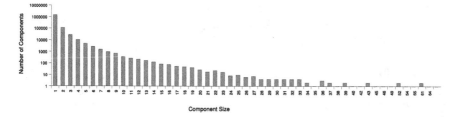

Fig. 3. Component size distribution in the query log sample

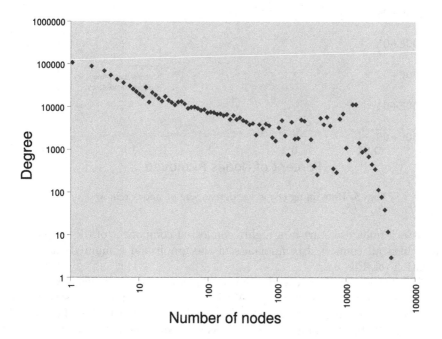

Fig. 4. Degree distribution in the largest graph component

contains approx. 3 million nodes and reflects a sample of the usage registered by the search engine in 2005. The original *Density* value for this graph is equal to 0.000089, which can already be considered low in comparison to the maximum *Density* = 1.

First, we computed the likelihood of finding edges, i.e. non-empty intersections of result-sets, among pairs of queries that share a term. This corresponds to a subgraph of particular interest, because queries sharing a term might be a possible target for an attack. The *Density* value for this subgraph is 0.000045 and thus lower than the *Density* value of the original graph. This means that queries sharing terms do not necessarily share more clicked results than the rest of the queries. Thus, an adversary would not have an advantage by focusing on this subgraph. Therefore, we continue our analysis with the original graph.

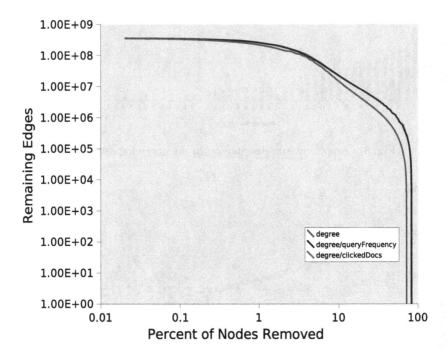

Fig. 5. Remaining edges vs. percentage of nodes removed

An attack may also start in a highly connected component of the graph. So, we identified all connected components of the graph and computed their size distribution (cf. Fig. 3).

We found that there is a very big connected component which includes almost 50% of all of the nodes in the graph. Without loss of generality, we study this big connected component hereafter. We first analyze the distribution of the node degrees in it. If this distribution corresponds to a *power law*, then, by removing the nodes with the highest degrees it would be possible to disconnect the graph very quickly. This would indicate that this query log is very likely to be success-fully anonymized with very little loss of information. However, as we can see in Fig. 4, the degree distribution is *not* a power law. It seems that approximately 9% of the nodes have a very high degree, so we cannot disconnect the graph by removing only a few high-degree nodes.

6.2   Three Methods for Graph Disconnection

We focus our study on the big connected component of the original graph. To disconnect it, we used the heuristic described in Section 5. We defined three variations of the heuristic by using different weighting schemes:

Method 1 (degree): The nodes are sorted only by their degree, the node weight defaults to 1.

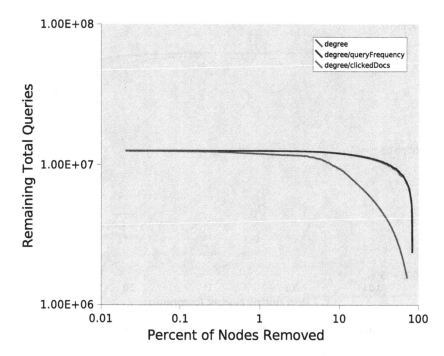

Fig. 6. Remaining volume of queries vs. percentage of nodes removed

Method 2 ($\frac{degree}{queryFrequency}$): The nodes are sorted by their degree divided with the *frequency of the query in the log.*

The *frequency* of a query (or node) is defined as the total number of times the query was submitted to the search engine.

Method 3 ($\frac{degree}{clickedDocs}$): The nodes are sorted by their degree divided with the *number of clicked documents for the query in the log.*

This number *clickedDocs* is the total number of times that documents were clicked from the result-set for that query.

6.3   Result Overview

Figures 5, 6 and 7 show the relation among removed nodes, remaining query volume and documents for each method. Each figure shows how the different log contents decrease until all edges have been removed from the query log graph. It can be seen that the number of nodes removed to disconnect the graph is very high for all methods. Nonetheless, the ultimate objective of retaining a large dataset is satisfied: The retained dataset still contains approx. 2,500,000 *total queries* (Fig. 6) and 1,200,000 *clicked documents* (Fig. 7). Here, *total queries* is the sum of the frequencies of the remaining nodes (queries), while *clicked documents* is the sum of all the clicks to documents from the remaining queries.

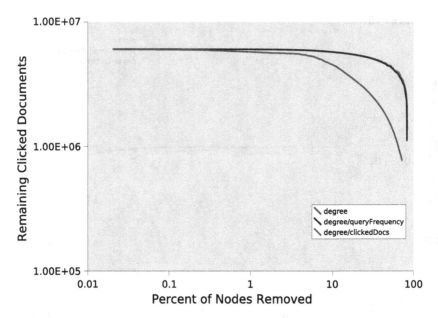

Fig. 7. Remaining volume of clicked documents vs. percentage of nodes removed

6.4   Comparing the Three Methods

The best-performing variation of the heuristic is Method 2, which sorts nodes on degree divided by query frequency: It removes less queries and clicked documents than Method 1. It dominates Method 3, which behaves similarly to Method 2 but eliminates more nodes. Thus, Method 2 retains the largest log volume.

Method 1 scans the log by processing the node with the highest degree first. When we compare the curves in the three figures, we see that the disconnection of the graph requires the removal of slightly less nodes (Fig. 5). However, the method removes a larger number of total queries (Fig. 6) and clicked documents (Fig. 7) than the other two methods. Therefore, it is inferior to the other methods with respect to the objective of retaining a large dataset.

In Fig. 6 and Fig. 7, we can see that Method 1 follows a different curve than the other two methods: The slope drops smoother and the drop starts earlier. This observation indicates that the connectivity of the graph drops earlier under this Method. We want to study extensions of this heuristic that will allow us to shift the end-point of the curve to the left, i.e. eliminate less nodes.

7   Conclusions and Future Work

In this paper we have presented a new issue on privacy preserving data analysis in the Web, the *preservation of website privacy*. We have shown that website information can be extracted from naively anonymized query logs. We have also defined different types of adversaries encountered when dealing with website information, as well as general types of vulnerabilities, which can be used to disclose information. We have presented specific attacks and techniques to prevent them.

We have described a graph representation for query log privacy preservation analysis and have defined a heuristic for log anonymization through graph disconnection. We have derived three methods upon this heuristic by considering different ways of sorting the nodes in the graph and then removing the highest rank nodes.

One of the methods, which sorts and the nodes (queries) on their degree divided by the query frequency, is experimentally shown to be the best in preserving the most amount of log volume, i.e. total number of queries and clicked documents. The complete disconnection of the graph requires removing most of the queries, but the statistical properties of the remaining ones still allow for knowledge discovery tasks. Also, the disconnection of the graph can be achieved by removing infrequent queries. Infrequent queries are those most likely to point to individuals (persons or institutions), so it is intuitively desirable to remove them.

The graph statistics described in [13] and the fact that query logs usually follow stable distributions indicate that the results obtained from this log can scale to logs of longer time periods and to query logs from other search engines. Queries that are removed by our anonymization technique are infrequent, minimizing the loss of potentially useful information in the remaining data.

Another important characteristic of the heuristic presented in this work is that the graph representation of the query log can be computed relatively fast. This makes the our anonymization approach suitable for rapidly changing data, such as query logs.

Future work on this topic includes studying the possibility of making infrequent queries frequent by generalization. This would imply replacing queries with their keywords. It is worth studying whether query generalization reduces the vulnerability posed by infrequent non-disconnected queries.

Acknowledgments

The authors thank Alessandro Tiberi from the University of Rome "La Sapienza" for providing the graph representation of the query log and help to understand this data. Also we thank the following people from Yahoo! Research: Aristides Gionis for many valuable discussions and feedback, and Vanessa Murdock and Bo Pang for their corrections for this final version.

References

1. AOL research website, no longer online, http://research.aol.com
2. Arrington, M.: AOL proudly releases massive amounts of private data (2006), http://www.techcrunch.com/2006/08/06/aol-proudly-releases-massive-amounts-of-user-search-data/

3. Barbaro, M., Zeller, T.: A face is exposed for AOL searcher no. 4417749, New York Times (2006)
4. Kumar, R., Novak, J., Pang, B., Tomkins, A.: On anonymizing query logs via token-based hashing. In: WWW 2007: Proceedings of the 16th international conference on World Wide Web, pp. 629–638. ACM Press, New York (2007)
5. Adar, E.: User 4xxxxx9: Anonymizing query logs. In: Query Log Analysis: Social and Technological Challenges, Workshop in WWW 2007 (2007)
6. Verykios, V., Bertino, E., Fovino, I., Provenza, L., Saygin, Y., Theodoridis, Y.: State-of-the-art in privacy preserving data mining. SIGMOD Record 33(1), 50–57 (2004)
7. Chawla, S., Dwork, C., McSherry, F., Smith, A., Wee, H.: Toward privacy in public databases. In: Theory of Cryptography Conference, pp. 363–385 (2005)
8. Kifer, D., Gehrke, J.: Injecting utility into anonymized datasets. In: Proceedings of the 2006 ACM SIGMOD international conference on Management of data, pp. 217–228 (2006)
9. Aggarwal, C., Pei, J., Zhang, B.: On privacy preservation against adversarial data mining. In: Proceedings of the 12th ACM SIGKDD international conference on Knowledge discovery and data mining, pp. 510–516 (2006)
10. Samarati, P., Sweeney, L.: Protecting privacy when disclosing information: k-anonymity and its enforcement through generalization and suppression. Technical report (1998)
11. Broder, A.: A taxonomy of web search. ACM SIGIR Forum 36(2), 3–10 (2002)
12. Albert, R., Jeong, H., Barabasi, A.L.: Error and attack tolerance of complex networks. Nature 406(6794), 378–382 (2000)
13. Baeza-Yates, R., Tiberi, A.: Extracting semantic relations from query logs. In: ACM SIGKDD international conference on Knowledge discovery and data mining (to appear, 2007)

Privacy-Preserving Data Mining through Knowledge Model Sharing*

Patrick Sharkey, Hongwei Tian, Weining Zhang, and Shouhuai Xu

Department of Computer Science, University of Texas at San Antonio
{psharkey,htian,wzhang,shxu}@cs.utsa.edu

Abstract. Privacy-preserving data mining (PPDM) is an important topic to both industry and academia. In general there are two approaches to tackling PPDM, one is statistics-based and the other is crypto-based. The statistics-based approach has the advantage of being efficient enough to deal with large volume of datasets. The basic idea underlying this approach is to let the data owners publish some sanitized versions of their data (e.g., via perturbation, generalization, or ℓ-diversification), which are then used for extracting useful knowledge models such as decision trees. In this paper, we present a new method for statistics-based PPDM. Our method differs from the existing ones because it lets the data owners share with each other the knowledge models extracted from their own private datasets, rather than to let the data owners publish any of their own private datasets (not even in any sanitized form). The knowledge models derived from the individual datasets are used to generate some pseudo-data that are then used for extracting the desired "global" knowledge models. While instrumental, there are some technical subtleties that need be carefully addressed. Specifically, we propose an algorithm for generating pseudo-data according to paths of a decision tree, a method for adapting anonymity measures of datasets to measure the privacy of decision trees, and an algorithm that prunes a decision tree to satisfy a given anonymity requirement. Through an empirical study, we show that predictive models learned using our method are significantly more accurate than those learned using the existing ℓ-diversity method in both centralized and distributed environments with different types of datasets, predictive models, and utility measures.

1 Introduction

Personal data collected by government, business, service, and educational organizations is routinely explored with data mining tools. Although data mining is typically performed within a single organization (data source), new applications in healthcare, medical research, fraud detection, decision making, national security, etc., also need to explore data over multiple autonomous data sources. A major barrier to such a distributed data mining is the concern of privacy: data owners must balance the desire to share useful data and the need to protect private information within the data.

---

* This work was supported in part by NSF reseach grant IIS-0524612.

F. Bonchi et al. (Eds.): PinKDD 2007, LNCS 4890, pp. 97–115, 2008.

Two approaches of privacy-preserving data mining (PPDM) can be identified in the literature: one is crypto-based and the other is statistics-based. The
crypto-based PPDM approach, such as [1,2], requires data owners to cooperatively execute specially designed data mining algorithms. These algorithms provide provable privacy protection and accurate data mining results, but often
suffer performance and scalability issues. On the other hand, statistics-based
PPDM approach, such as [3,4,5,6,7,8,9,10], lets each data owner release a sanitized dataset (e.g., via perturbation [3] or generalization [10]) to a third party,
who can then execute data mining algorithms to explore the published data.
Statistics-based methods have efficient implementations, making them more appropriate than crypto-bases methods to deal with large volumes of data.

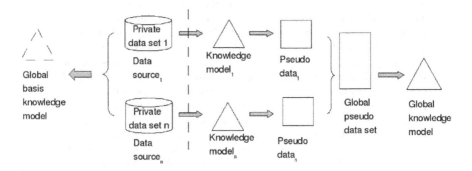

Fig. 1. Privacy-preserving data mining by knowledge model sharing: the goal is that
the two resulting global knowledge models are (almost) the same

One problem of statistic-based data publishing methods is that they suffer loss
of data quality, probably because they treat data sanitizing and data mining as
unrelated tasks. For example, suppose a data miner wants to learn a decision
tree classifier from the published data. Since decision tree learning algorithms
select attributes for tree node split according to data distributions, in order for
these algorithms to learn high quality decision trees from a published dataset,
the sanitization methods should preserve the distributions of best splitting attributes. But, since data sanitizing and data mining are not considered together
in a data publishing framework, the sanitization methods often indiscriminately
alter data values, causing changes to the distributions of those important attributes, thus adversely affect the accuracy of classifiers learned from sanitized
data. Even if there are opportunities to satisfy a privacy requirement and to preserve data quality by modifying data of some less important attributes, existing
data publishing methods are unable to take advantage of those opportunities.

To address this problem, we introduce in this paper a new approach: *knowledge
model sharing* (or simply model sharing), which lets each data source release a
privacy-preserving local knowledge model learned from its private data, and let
a data miner explore pseudo data generated from the local knowledge models.
Specifically, as indicated in Figure 1, each data source learns a type of knowledge

model closely related to what a data miner wants to learn, using conventional data mining algorithms, and modifies the model according to a privacy requirement before releasing it to the data miner. The data miner uses local knowledge models to generate local pseudo datasets and combines them into a single dataset before applying conventional data mining algorithms to learn global knowledge models. In this paper, we focus on using predictive models as global knowledge models, with decision trees or classification rule sets as local knowledge models.

By not releasing any data (not even a sanitized version) and by modifying local knowledge models, model sharing can effectively protect privacy. In addition, as our study shows, this approach obtains much better quality data than data publishing methods do. Model sharing can be easily implemented since for example, data sources can easily offer data mining services [11,12] using widely available data mining tools and service-oriented computing techniques.

While model sharing is instrumental, a number of technical subtleties need to be carefully addressed. In this paper, we make the following specific contributions.

1. For data miners, we define the problem of generating pseudo-data from a predictive model and present an efficient heuristic algorithm that generates a pseudo dataset from a decision tree, according to its paths. This algorithm can be easily extended to generate pseudo datasets from disjoint classification rules.

2. To measure privacy of a predictive model, we show how to adapt privacy measures of data, such as k-anonymity and ℓ-diversity, to measure the privacy of decision trees whose class labels represent sensitive information. To our knowledge, privacy measure of knowledge models has not been proposed in the literature.

3. To protect private information contained in a predictive model, we present an algorithm for a data owner to prune a decision tree according to a given anonymity measure. This tree pruning technique can be viewed as a special form of generalization that preserves important patterns in the raw data. Used together with a good pseudo data generation method, this technique can result in higher data quality, compared to data publishing methods.

4. We perform an empirical study of the pseudo-data generation and tree pruning techniques. Specifically, we measure the utility of the pseudo data by the quality, such as classification accuracy, of global predictive models. For pseudo-data generation, we compare predictive models learned from global pseudo datasets with those learned directly from the local raw data. For tree pruning, we compare predictive models learned respectively from raw data, pseudo data, and ℓ-diversified data. In addition to the decision tree, we also study other types of global predictive models, such as Naive Bayes and conjunctive rules. Our results show that predictive models learned using our method are significantly more accurate than those learned using the ℓ-diversity method in both centralized and distributed environments with different types of datasets, predictive models, and utility measures.

We focus on data quality and limit the comparison of our tree pruning technique to the ℓ-diversity technique because, while data mining provides a framework for comparing data utility of different privacy-preserving techniques, no metric currently exists for comparing privacy assurance of these techniques, making it impossible to fairly compare different privacy protection methods. We are able to compare pseudo data produced by tree pruning method with ℓ-diversified data only because the ℓ-diversity measure of data and the ℓ-diversity measure of decision trees are very similar. Finding a metric that can be used to compare privacy assurances of all the PPDM techniques is still an open problem.

The rest of this paper is organized as follows. In Section 2, we define the problem of pseudo-data generation from a decision tree and present a path-based pseudo-data generation algorithm for data miners. In Section 3, we present the l-diversity privacy measure for decision tree and an algorithm that prunes a decision tree to satisfy a given anonymity-based privacy requirement for data owners. In Section 4, we describe our empirical study and present experiment results. In Section 5, we discuss previous work related to our work presented in this paper. We draw conclusions in Section 6.

2   Pseudo-data Generation from Decision Trees

We consider n data sources whose private data tables (or datasets) form a horizontal partition[1]. We assume that private data tables contain no personal identifier, such as name or identity number, but may contain duplicate tuples.

The goal of a data miner is to learn a global decision tree (see Figure 1) from a collection of local pseudo tables. In order to use this global decision tree to classify unseen tuples, local pseudo tables must preserve sufficient information contained in local private tables. One way to measure the quality of local pseudo tables is to compare the quality of the global decision tree to that of the global basis decision tree, learned directly from the (unavailable) collection of private tables (see Figure 1). Here, instead of seeking for high absolute quality, we are looking for high relative quality: the quality of the global (learned) decision tree should be as high as that of the global basis decision tree. Thus, one may want to formulate the problem as to find optimal local pseudo tables that guarantee identical quality of the global learned and basis decision trees. But unfortunately, this problem is ill-defined because in practice the global basis decision tree is not available to the data miner.

To overcome this difficulty, we focus on the problem of generating a high quality pseudo table for a single data source. Once this problem is solved, we can construct a high quality global pseudo table by combining high quality local pseudo tables. With this in mind, we define the pseudo-data generation problem as follows.

---

[1] In a distributed environment, private data tables of data sources can form a horizontal, vertical, or hybrid partition, where tables of a horizontal partition have a common set of attributes and those of a vertical partition have different sets of attributes.

Problem Statement: Find a pseudo dataset D based on a given basis decision tree h so that it minimizes the difference of classification accuracies, measured using any testing set, between h and the decision tree h' learned from D.

Notice that the difficulty involving multiple data sources is avoided here because both the basis decision tree h and the learned decision tree h' are available: h is published by the data source and h' is learned from a pseudo table.

This problem has at least one solution: the private table used to learn the basis tree h. Obviously, if a pseudo table is the same as the private table, the learned and the basis trees can have identical structures, and therefore, identical classification accuracies. However, we are more interested in other solutions to this problem (for an obvious reason).

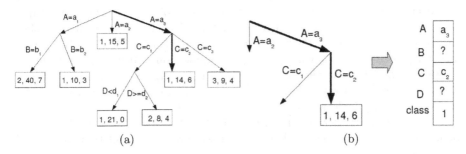

(a)   (b)

Fig. 2. A sample decision tree with leaf labeled by (class label, hit count, miss count)

Finding an optimal pseudo table from a given basis decision tree is difficult for several reasons. First of all, due to inherent randomness of decision tree algorithms (for example, they make random choices among several possible split attributes of identical class purity), different decision trees may be learned from the same dataset. Thus, even if a pseudo table is indeed the private table, there is no guarantee that the learned and the basis trees will have identical structures. Secondly, if two decision trees have different structures, it is difficult to determine if the difference of their classification accuracies is minimal.

In the rest of this section, we present a heuristic algorithm for a data miner to generate pseudo data from a decision tree. This algorithm uses information contained in a typical decision tree, as we shall describe next.

2.1   Decision Tree and Pseudo-data

A decision tree (see Figure 2(a) for an example) consists of labeled nodes and edges. The label of an edge is a predicate $A = a$ specifying that attribute A (of a tuple) has a value a, where a is a simple value if A is categorical and an interval if A is numeric. All the outgoing edges of a node are labeled by the same attribute and the values in these labels partition the domain of the attribute.

Each leaf node v is labeled by a class $v.class$, the majority (or plurality) class of training tuples assigned to the node during decision tree learning, and two counts: a hit count $v.hit$ representing the number of training tuples[2] in the majority class at this node, and a miss count $v.miss$ representing the number of misclassified training tuples (in other classes) at the node.

Each root-to-leaf path p of the tree has a path label $label(p)$, a class label $class(p)$, a hit count $hit(p)$ and a miss count $miss(p)$, where $label(p)$ is a conjunction of edge labels, the class label and the two counts are those of the leaf node of the path. In the path label, each attribute appears at most once and if $A = a$ appears, a is the most specific value of A in the entire path. We denote the set of attributes of path p by $attr(p)$, and the value of an attribute $A \in attr(p)$ by $value(A, p)$. For example, let p be the second leftmost path in the decision tree of Figure 2(a). Then, $label(p) = (A = a_1 \wedge B = b_2)$, $class(p) = 1$, $hit(p) = 10$, $miss(p) = 3$, $attr(p) = \{A, B\}$, and $value(A, p) = a_1$.

A decision tree can be used to predict the class of unseen tuples. If a tuple t satisfies a path p of a decision tree, that is, if each attribute of t satisfies the condition specified in the path label, or equivalently $\forall A \in attr(p)\ t[A] \in value(A, p)$, the class of t will be $class(p)$. In addition, a decision tree is also a descriptive model. In particular, the decision tree local to a data source can be viewed as an abstract representation of the local private table, providing a basis for generating a pseudo table.

By considering the learned the the basis decision trees that are structurally identical, we obtain following characterization of an important type of optimal pseudo tables.

Proposition 1. *Let D be a pseudo table generated from a given basis decision tree h, and h' be a decision tree learned from D. If h and h' have identical structure, that is $h = h'$, then every tuple in D satisfies a path of h. Furthermore, for each path $p \in h$, D contains exactly $hit(p) + miss(p)$ tuples that satisfy p.*

Proof. Since h and h' have identical structure, every path of h' is also a path of h, that is $\forall p(p \in paths(h') \iff p \in paths(h))$. According to the information contained in a decision tree, since D is the training set of h', each tuple of D must satisfy exactly one path of h', therefore, it also satisfies the same path of h. Consequently, the hit and miss counts of each path in h' report exactly the number of tuples in D that satisfy the path.

Although the properties described in Proposition 1 are not necessary, because not all optimal pseudo datasets result in identically structured learned and basis trees, they provide a basis for generating pseudo tuples: paths can define templates of pseudo tuples. For example, the label of the path in Figure 2(a), highlighted by thick lines, defines the template tuple in Figure 2(b), where values of attributes A and C are specified by the path label, and values of attributes B and D, absent from the path, are yet to be determined (indicated by question marks).

---

[2] To ease the presentation, we consider only the actual counts. It is straightforward to extend this to relative frequencies.

2.2   Path-Based Data Generation

It is difficult to determine optimal values of a template tuple for attributes absent from a given path. Let us consider a significantly simplified case. Assume that a decision tree contains s paths, where each path has r edges labeled with distinct attributes and is responsible of generating k pseudo tuples. Suppose each tuple has m attributes and each attribute has exactly K distinct values. Since a pseudo table may contain duplicate tuples, the k pseudo tuples of a path can be generated independently. Similarly, since paths of the decision tree partition tuples into disjoint groups, pseudo tuples of different paths can also be generated independently. Therefore, there are $N = ((K^{m-r})^k)^s$ possible pseudo datasets that satisfy the properties stated in Proposition 1. To the best of our knowledge, there is no known efficient algorithm for finding an optimal pseudo table using template tuples defined by decision tree paths. An exhaustive search is obviously too expensive.

We solve the problem with a simple heuristic that exploits impurity measures (such as information gain, Gini-index, or miss-classification rate) used by decision tree learning algorithms. We view the generation of a pseudo table as an inverse process of decision tree learning. Since at each tree node, the attribute that leads to the highest gain of class purity, based on a given impurity measure, is chosen by the learning algorithms to split the training tuples in the node, a pseudo-data generation algorithm can try to force the learning algorithm to select the same attribute, by making sure that the class purity of non-split attributes in the pseudo data is lower than that of the split attribute. Although distributions of non-split attributes in raw data are unknown, we can still minimize the class purity of these attributes in pseudo data by assigning random values from a uniform distribution.

Based on this heuristic, algorithm PGEN (in Figure 3) uses each path of a decision tree to generate a set of (not necessarily distinct) tuples. In line 2 of the algorithm, each path is considered independently. The number of tuples to be generated by a path (see line 3) can be determined in several ways. For example, it can be the sum of the hit and miss counts, or a proportion of a user specified table size.

In line 4 of the algorithm, an empty template tuple is generated and an un-specified value is assigned to each attribute. Subsequently, for each attribute that appears in the path, a value is assigned by function $value(A, p)$ according to the type of the attribute: if the attribute is categorical, the value is the one specified in the path label (line 7); if the attribute is numeric, the value is chosen within the interval specified in the path label (line 8), which can be a fixed default value, such as the middle point or an end point of the interval, or a random value drawn from the interval according to a given distribution. For each attribute that is absent from the path, a random value from its domain is assigned (line 9). Finally, in line 10, a class label is assigned according to the distribution of classes indicated by the hit and miss counts of the path. Non-majority classes can be randomly assigned to the tuple according to a known distribution or a uniform distribution, the latter is used in our experiments.

Algorithm Path-based Pseudo-data Generation (PGEN)
Input: A decision tree T and a set of attributes $\mathcal{A}$ of data
Output: A set S of labeled tuples in $\mathcal{A}$
Method:

1. $S = \emptyset$;
2.   for each path p of T do
3.   for pseudo tuple t to be generated by p do
4. $t = $ a new empty tuple;
5.   for each attribute $A \in \mathcal{A}$ do
6.   if $A \in attr(p)$ then
7.   if A is categorical then $t[A] = value(A, p)$;
8.   else $t[A] = $ a value selected from $value(A, p)$);
9.   else $t[A] = $ a value selected from the domain of A;
10. $t[class] = $ a class selected based on counts of p;
11.   append t to S;
12.   return S;

Fig. 3. Path-based Pseudo-data Generation Algorithm

3   Privacy-Preserving Tree Pruning

So far, we only considered the pseudo-data generation by the data miner. We now consider what a data owner needs to do. If the local predictive model of a data source does not leak private information, the data owner needs to do nothing more than releasing the local model. However, a predictive model may leak private information. For example, assume that the class labels of leaf nodes in the decision tree in Figure 2(a) are values of a new attribute, say SA, referred to here as a sensitive attribute, representing some private information. Then the path labeled by $A = a_3, C = c_1, D < d_1$ is open to the homogeneous attack described in [9], because its hit and miss counts reveal that all the tuples in that path have the same sensitive value $SA = 1$.

Although it is possible for a decision tree learned from a private table to include no sensitive attribute, hence to leak no private information, it is common that a decision tree has a sensitive attribute as its class, just like the tree in Figure 2(a). Thus it is important to prevent a knowledge model from leaking private information. A critical question is how to measure the privacy of a knowledge model. In the following, we show that anonymity measures of dataset, such as ℓ-diversity, can be easily adapted to measure privacy of decision trees. In doing so, the privacy of published data and that of pseudo data can be compared with each other.

3.1   Anonymity Measures of Decision Trees

Interestingly enough, there is a close analogy between paths of a decision tree learned from private tables and equivalence classes in an anonymous dataset produced by generalization methods, such as k-anonymity and ℓ-diversity. In generalization methods, tuples in a private table are generalized to satisfy a privacy requirement according to taxonomies of quasi-identifier (QI) attributes. The generalized tuples form equivalence classes, called QI-groups, in which all tuples have the same QI value. Each QI-group in a generalized dataset needs to satisfy a privacy requirement, such as containing at least k tuples (under k-anonymity) or having well-represented sensitive values (under ℓ-diversity). On the other hand, paths in a decision tree are similar to QI-groups of a generalized dataset. Like QI-groups, paths also partition tuples into equivalence classes: all the tuples in one path have the same values under attributes that are present in the path. We can adapt anonymity measures of QI-groups to measure the privacy of decision tree paths. For example, to measure paths with k-anonymity, we can require each path p of the decision tree to satisfy $p.hit + p.miss \geq k$.

Similarly, we can also measure paths with an ℓ-diversity measure. For example, one of the ℓ-diversity measures proposed in [9] is (c, ℓ)-diversity, which requires every QI-group to satisfy $d_1 \leq c \cdot \sum_{i=\ell-1}^{m} d_i$, where d_i, $1 \leq i \leq m$, is the number of tuples of the ith largest class within the QI-group. To apply (c, ℓ)-diversity to measure a decision tree, we can require every path p to satisfy $\ell \leq p.miss + 1$ and $p.hit < c \times \frac{p.miss}{\ell-1}$. Intuitively, the largest class of the path, $p.class$, contains $p.hit$ tuples, and all the other classes collectively contain $p.miss$ tuples. If $p.miss > 0$, we need to estimate the sizes of non-majority classes. Specifically, if $p.miss < \ell - 1$, there are at most $\ell - 2$ other classes (each contains one tuple), therefore, the path trivially violates ℓ-diversity. If $p.miss \geq \ell-1$, there may be ℓ or more classes in the path. Since the class distribution of misclassified tuples is not provided by the decision tree, the adversary may have to assume that misclassified tuples are equally distributed over $\ell - 1$ non-majority classes, with $\frac{p.miss}{\ell-1}$ tuples per class, with the smallest class representing all remaining classes (if there are more than ℓ classes. For example, the highlighted path in Figure 2(a) indicates that there are 14 tuples in class 1, and 6 tuples in classes 2 and 3 (assuming there are total three classes), with three tuples in each class. As a result, this path is $(5, 3)$-diversified.

3.2   Anonymity-Based Tree Pruning

With generalization methods, QI-groups that does not satisfy a user-specified anonymity requirement will be combined with other QI-groups. The corresponding operation for a decision tree is to combine paths of a common prefix into a single path, which effectively prunes the decision tree. We now present a simple algorithm that prunes a decision tree based on a given anonymity requirement (such as k-anonymity or (c, ℓ)-diversity).

Algorithm APT (see Figure 4) makes two traversals of the tree: a depth-first traversal (lines 2-4) that propagates class and hit/miss counts from leaf nodes to their ancestor nodes, and a bottom-up level-by-level traversal (lines 6-14) that prunes the nodes that does not satisfy a given anonymity requirement. A node is pruned by being combined with some of its sibling nodes (lines 10-13). Once all children nodes of a parent node are pruned, the parent node is transformed

Algorithm Anonymity-based Tree Pruning (ATP)
Input: a decision tree T and an anonymity requirement a
Output: a tree T' that satisfies a
Method:

1. $T' = T$;
2.   for each node v encountered in a depth-first traversal of T' do
3.   if v is a leaf node then check(v,a);
4.   else compute $v.hit$, $v.miss$ and $v.class$;
5. $L =$ the maximum level of T';
6.   while $|T'| > 2$ and not all leaf nodes marked pass do
7. $L = L - 1$;
8.   for each non-leaf node v at level L do
9.   while v has a bad child b and a merge is possible do
10.   merge b with suitable siblings into a new child u of v;
11.   check(u,a);
12.   if v has a single child u then
13.   replace v by u;
14.   else mark v pass;
15.   if $|T'| > 2$ return T' else return an empty tree;

Fig. 4. Anonymity-based Tree Pruning Algorithm

into a new leaf node. If the algorithm terminates successfully, it will produce a decision tree satisfying the anonymity requirement. Otherwise, it produces an empty tree.

During the depth-first traversal, all the leaf nodes are checked against the anonymity requirement, and in the meantime, the hit and miss counts as well as class are calculated for each non-leaf node. A node is marked if it is a leaf node and satisfies the anonymity requirement (Line 3) or all its children are marked (Line 13). In Line 4, the hit and miss counts, as well as the label of the majority class are calculated in a bottom-up fashion. Assume a non-leaf node has k children and the set of classes is C. The majority class of the non-leaf node is the class containing the most tuples in its sub-tree. Due to the recursive traversal, the number of tuples in each class c_j in a node can be estimated as follows, using the hit and miss counts of the node's children nodes:

$$hit_{c_j} = \sum_{v.class=c_j} v.hit + \frac{1}{|C|-1} \sum_{v.class \neq c_j} v.miss \qquad (1)$$

where $\frac{v.miss}{|C|-1}$ is the expected number of misclassified tuples in v that belongs to class c_j. Intuitively, if the majority class of a child node is c_j, this child node will contribute $v.hit$ tuples to class c_j of its parent node. If the majority class of a child node is not c_j, it will still contribute a portion of $v.miss$ tuples to class c_j of its parent node. To determine the size of that portion, we assume that all non-majority classes are equally likely among misclassified tuples. This can be extended easily to use actual distributions of classes in children nodes, if that information is available. Once the hit counts of class labels are calculated, the class and counts of the parent v can be obtained as the following: $v.class = c = \max\arg_{c_j \in C}\{hit_{c_j}\}$, $v.hit = hit_c$, and $v.miss = \sum_{c \in C}(hit_c + miss_c) - v.hit$.

During the bottom-up traversal, starting from the second-to-the-bottom level, the algorithm looks for nodes whose children do not all satisfy the anonymity requirement. Once found, the violating node will be merged with some of its siblings (lines 8-10). Suppose b is the violating node and it has an incoming edge labeled by $A = a$. Different criteria can be used to select the siblings to be merged with b. For example, we can merge all the siblings of b. Alternatively, we can merge b with those siblings whose incoming label is $A = a'$ where a' is a descendant of a. To merge a set of siblings, we remove them together with all their descendants, and then add a single new node under their parent. The hit count, miss count, and class label of this new node are determined in the same way as in line 4, treating the set of sibling nodes as the children of the new node.

If the pruning terminates successfully, the algorithm returns a tree containing a root and at least two leaf nodes, that is $|T'| > 2$. Otherwise, it simply returns an empty tree.

4   Empirical Study

We evaluate the path-based pseudo-data generation and anonymity-based decision tree pruning techniques empirically by comparing the quality of predictive models learned from the pseudo-data with the quality of those models learned directly from the private data. We also compare these techniques with the ℓ-diversity technique of data publishing.

We implemented the PGEN algorithm (described in Section 2.2) and the ATP algorithm (described in Section 3.2) in Java and performed extensive experiments on a Pentium PC with 2GB memory.

The five datasets listed in Table 1, from the UCI Machine Learning Repository, were used in our study. For the purpose of the study, these datasets were preprocessed to remove missing values.

In our experiments, decision trees are always used as local predictive models, but different models, including Naive Bayes classifiers, conjunctive rules, decision

Table 1. Experimental datasets

datasets	#Instances	# Attributes		#Classes
		Nominal	Numeric	
Adult	45204	9	6	2
Car	1728	7	0	4
CMC	1473	8	2	3
Nursery	12960	9	0	5
Train	3000	9	1	2

tables, random forests and decision trees, are used as the global model. These models are considered because they are widely used and are able to handle our datasets. In our study, these predictive models are learned using public-domain Java implementations of their respective learning algorithms.

We consider two pseudo-data generation methods. One method, referred to as BASEGEN, uses the published decision tree to predict and label tuples randomly selected from the data space. The other is PGEN, the path-based pseudo-data generation method.

The quality of predicative models is measured by classification accuracy and class match, defined respectively as the percentage of testing tuples that are correctly classified by the predictive models being tested and the percentage of testing tuples that are classified identically by the models being compared. In addition to these two measures, for decision trees, we also measure path match, defined as the percentage of paths shared by the decision trees under comparison. That is,

$$ss(h, h') = \frac{|paths(h) \cap paths(h')|}{|paths(h) \cup paths(h')|}$$

where for any decision tree h, $paths(h)$ denotes the set of root-to-leaf paths of h.

To account for the randomness in the data, all the quality measures are obtained using the ten-fold validation method and reported here as the average over five iterations. To evaluate the impact of data distribution, we consider distributed environments with 1 to 10 data sources. Specifically, in each run, one fold of data is reserved as the global testing set. The other nine folds are first used to learn a global basis predictive model, referred to as BASIS, and then are randomly partitioned into equal-sized private tables among all the data sources. All the data sources use the same data generation algorithm to create pseudo-data.

To evaluate the tree pruning method, we compare two methods, one method, referred to as PPGEN, is ATP followed by PGEN, and the other method, referred to as LD, is the ℓ-diversity data publishing method. Specifically, we generate a ℓ-diversified dataset using LD and a pseudo dataset using PPGEN from the same raw data, and compare quality measures of predictive models learned from these datasets. Our Java-based implementation of LD is based on [9].

4.1   The Quality of Pseudo-data Generation

In this study, we compare the quality of the two pseudo-data generation methods: BASEGEN and PGEN. To focus on data generation, the decision trees used to generate pseudo-data are not pruned in this study.

Figure 5 shows accuracies of decision trees and Naive Bayes classifiers on three UCI datasets. In the two charts in this Figure, the X-axis represents the number of data sources, ranging from 1 to 10 (showing 1 to 5 in order for the graph to be readable) and the Y-axis is the average classification accuracy. The group of 9 bars consists of three subgroups, each having 3 bars, corresponding to

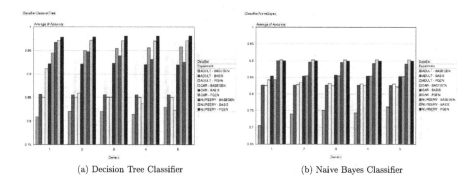

(a) Decision Tree Classifier   (b) Naive Bayes Classifier

Fig. 5. Accuracy of a single predictive model on Adult, Car and Nursery datasets

BASEGEN, BASIS, and PGEN results with Adult, Car, and Nursery datasets, respectively. The accuracies of BASIS classifiers stay the same across all groups. The results on the other two UCI datasets are similar to those presented here and are omitted to save space. In Figure 5(a), PGEN performs very well as compared to BASIS. BASEGEN on the other hand, consistently performs worse than BASIS. It is interesting that for Car and Nursery datasets, as the number of owners increases, the accuracy of PGEN increases and that of BASEGEN decreases. This trend is not observed with Adult dataset. We observe a similar trend in Figure 5(b), which uses Naive Bayes classifiers rather than decision trees as the basis and learned models, although the difference between BASEGEN and PGEN are not as big as in Figure 5(a) for Car and Nursery data. An exception is with the Adult data, where as the number of data sources increases, the accuracy of BASEGEN improves more significantly than in in Figure 5(a). Similar trends are also observed in results with other predictive models and datasets. From these results, we can conclude that PGEN can generate high quality pseudo-data that effectively preserves the information in the raw data, and BASEGEN can also be useful for some datasets with a small number of data sources.

Figure 6 shows classification accuracies of Conjunctive Rule, Decision Tree and Naive Bayes models with (a) Adult data and (b) Nursery data. Both Decision Tree and Naive Bayes models perform better on Nursery data than on Adult data. The Conjunctive Rule does just the opposite. Accuracy of global decision trees are better than the other two models, which is expected because published local models are also decision trees. It is interesting however that the Naive Bayes model performs equally well with both BASEGEN and PGEN on Nursery data, but does better with PGEN than with BASEGEN on Adult data. Also notice that the conjunctive rule model performs much better with BASEGEN than it does with PGEN on Adult data, which is somewhat surprising. The results obtained with other predictive models on other datasets are similar to those presented in Figure 6(b).

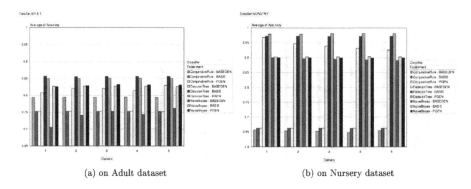

(a) on Adult dataset   (b) on Nursery dataset

Fig. 6. Accuracy of Conjunctive Rule, Decision Tree, and Naive Bayes classifiers

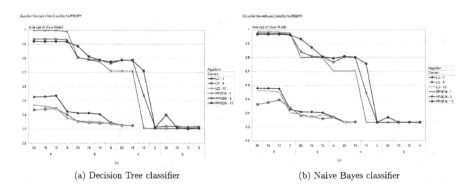

(a) Decision Tree classifier   (b) Naive Bayes classifier

Fig. 7. Privacy vs accuracy of classifiers on Nursery dataset

In addition to classification accuracy, we also obtained results on class match and path match. While the results on class match are very similar to that of classification accuracy, path match results reveal very little similarity between decision trees learned from raw data and those learned from pseudo-data, with typically less than 30% and frequently less than 1% shared paths. This result seems to suggest that the quality of pseudo datasets are not related to the structural similarity of decision trees.

4.2 Effect of Anonymity-Based Decision Tree Pruning

To evaluate ATP method, we compare the quality of PPGEN and LD over a range of (c, ℓ)-diversity privacy requirements, with ℓ ranging from 2 to 5 and c ranging from 5 to 20 (with an increment of 5). In out experiments, we generate an ℓ-diversified dataset using the LD method and learn a global basis predictive model (referred to as BASIS) from each global training set before it is partitioned among data sources. A global predictive model (also referred to as LD) is also learned from the ℓ-diversified dataset. To create a global pseudo dataset, the global training set is first partitioned among data sources, and subsequently used to learning local decision trees for data sources. These local decision trees are pruned using ATP under the same privacy requirements as used by LD to generate the l-diversified data. Then pseudo datasets are generated from local pruned decision trees using PGEN method, and subsequently combined into the global pseudo dataset. Finally, the global predictive models corresponding to PPGEN are learned. The classification accuracy as well as class match of the BASIS, LD, and PPGEN predictive models are compared here.

Figure 7 shows classification accuracy of decision tree and Naive Bayes models learned from the Nursery dataset. This result is representative of similar results obtained with other datasets. In Figure 7, the X-axis represents privacy requirements roughly ordered from left to right according to increasingly stronger privacy requirements, and the Y-axis represents the average accuracy. The lines represent various combinations of data generation methods (LD vs. PPGEN) and number of data sources (ranging from 1 to 10 with increment of 5). Again, we have to omit some results to make these figures legible.

From Figure 7 (a) and (b), we can see that PPGEN significantly outperforms LD (by more than 30 percentage points) over the entire range of privacy requirements that are tested. Furthermore, PPGEN continues to produce usable pseudo-data (with a 30% accuracy rate) as LD stops to produce any data due to exceedingly high privacy requirement (starting with $l = 4$ and $c = 10$). It is also interesting to see that the performance of PPGEN also falls sharply after $\ell = 4$ and $c = 10$, and that the increasing number of data sources seems to have a diminishing effect on accuracy.

These results may be attributed to multiple factors. First, LD performs a full dataset generalization on all QI-attributes, but PPGEN supports a hybrid generalization: different paths generalize their own attribute to their own levels, guided by the published decision tree. Second, since LD is applied to the private dataset, it suffers if the dataset is highly skewed or if it lacks a good distribution of classes, but PPGEN is better protected from this because it assumes an even distribution of misclassified tuples at every leaf node. This feature also explains why PPGEN can satisfy higher privacy requirements than LD. Finally, the results support the expected deterioration in quality as privacy increases. Similar results were obtained with the Car dataset.

Notice that, the pseudo dataset generated by PPGEN under a given pair of (l, c) is likely not to satisfy (c, l)-diversity on data. This is because the paths pruned by PPGEN differ from the QI-groups generalized by LD in that they typically involves fewer attributes than do QI-groups. Thus, tuples in the same path may belong to different QI-groups when measured using anonymity measures of datasets. Thus, a pseudo table may have many more smaller QI-groups than the corresponding ℓ-diversified dataset, leading to a violation of the data-oriented ℓ-diversity requirement. This however does not reduce the privacy assurance of PPGEN because the pseudo tuples are randomly generated rather than generalized from the raw data.

5   Related Work

In this section we briefly discuss the relationship between the work described in this paper and the work in several closely related areas. Our work is motivated by the problem of low data utility of existing methods that use perturbation or generalization to protect privacy.

Privacy-preserving data mining based on perturbation techniques was pioneered in [3], which perturbs data values by adding a noise and learns decision trees from an estimation of original data distribution. The adding-noise perturbation method was shown to be flawed in [13], and a number of other perturbation methods have been proposed including matrix perturbation [6,5], ρ_1-to-ρ_2 privacy breaching [4], and personalized breaching probability [14]. In addition to decision trees, perturbation have also been used in PPDM to find association rules [6,4,15,16] and clustering [17].

The well-known generalization method, K-anonymity, was first studied in [10] for data disclosure. Implementations of k-anonymity include bottom-up generalization [18], top-down specialization [19], and Incognito [20]. The k-anonymity is extended to ℓ-diversity in [9] to respond to two types of attacks against k-anonymity: homogeneous attack and background attack. To improve utility of ℓ-diversified data, [21] presented a method that publishes marginal tables in addition to ℓ-diversified tables. Another extension of k-anonymity, (α, k)-anonymity, is proposed in [22]. Because of the analogy of paths of a decision tree and QI-groups in data produced by these generalization methods, our tree pruning method can be used with any anonymity measure used by these methods.

Protecting privacy by generating pseudo data according to statistics of the private data has been proposed before. The method described in [23] generates pseudo data from statistics of condensed groups, which are sets of predetermined number of data records closest, in a multidimensional space, to randomly selected private records. The method in [7] learns a given number of clusters of at least a given size, and for each cluster, generate pseudo data according to the QI-values of the center, the size of the cluster, and the set of sensitive values in the cluster. Our method of generating pseudo data is different from these methods in that we use the statistics contained in decision trees only for attributes in their paths.

In conventional distributed data mining literature, there has been much work [24,25] on combining classifiers learned from different sources into a meta-classifier. These methods do not teak privacy into consideration and need to resolve conflict among different classifiers using a global testing set, which in our environment will contain private tuples. Our method learns global model from pseudo data rather than by integrating local models directly. This not only avoids using private tuples for testing, but also allows us to learn global models of different types.

The recent work on secure sharing of association rules and frequent patterns [26,27] focus on hiding sensitive association rules and blocking inference channels for the protection of privacy. However, in these papers, what considered private is some association rules or patterns rather than information about individuals.

Crypto-based methods, such as [1,2], do not have the privacy issue of local knowledge models because nothing is shared among data sources. But, in addition to performance and scalability issues, these methods still face the privacy issue of global knowledge models they produce, to which our method for measuring and preserving the privacy of knowledge models remains relevant.

6   Conclusions

In this paper, we present a new PPDM approach that aims to learn high quality knowledge models yet still protect privacy. With this knowledge model sharing approach, each data source releases a privacy-preserving local knowledge model learned from its private data, and a data miner mines pseudo data generated from the local knowledge models. We define the problem of generating pseudo-data from predictive models, such as decision trees and sets of classification rules, and present a heuristic algorithm for a data miner to generate pseudo-data according to paths of a decision tree. For data owners, we show how to adapt anonymity measures of datasets to measure the privacy of decision trees, and present an algorithm that prunes a decision tree to satisfy a given anonymity requirement. Our results, obtained through an empirical study, show that predictive models learned using our method are significantly more accurate than those learned using the existing ℓ-diversity method in both centralized and distributed environments with different types of datasets, predictive models, and utility measures.

Acknowledgment

The authors would like to thank the anonymous reviewers for their constructive comments that improved the quality of this paper.

References

1. Lindell, Y., Pinkas, B.: Privacy Preserving Data Mining. In: Bellare, M. (ed.) CRYPTO 2000. LNCS, vol. 1880, pp. 36–54. Springer, Heidelberg (2000)
2. Pinkas, B.: Cryptographic techniques for privacy-preserving data mining. ACM SIGKDD Explorations 4(2), 12–19 (2003)
3. Agrawal, R., Srikant, R.: Privacy-preserving data mining. In: ACM SIGMOD International Conference on Management of Data, pp. 439–450. ACM, New York (2000)
4. Evfimievski, A., Gehrke, J., Srikant, R.: Limiting privacy breaching in privacy preserving data mining. In: ACM Symposium on Principles of Database Systems, pp. 211–222. ACM, New York (2003)
5. Dowd, J., Xu, S., Zhang, W.: Privacy-preserving decision tree mining based on random substitutions. In: International Conference on Emerging Trends in Information and Communication Security, Freiburg, Germany (June 2006)

6. Agrawal, S., Haritsa, J.R.: A framework for high-accuracy privacy-preserving mining. In: IEEE International Conference on Data Engineering (2005)
7. Aggarwal, G., Feder, T., Kenthapadi, K., Khuller, S., Panigrahy, R., Thomas, D., Zhu, A.: Achieving anonymity via clustering. In: Proceeding of the 25th ACM Symposium on Principles of Database Systems (June 2006)
8. Aggarwal, C.C.: On k-anonymity and the curse of dimensionality. In: International Conference on Very Large Data Bases, pp. 901–909 (2005)
9. Machanavajjhala, A., Gehrke, J., Kifer, D., Venkitasubramaniam, M.: ℓ-diversity: Privacy beyond k-anonymity. In: IEEE International Conference on Data Engineering (2006)
10. Samarati, P., Sweeney, L.: Protecting privacy when disclosing information: k-anonymity and its enforcement through generalization and suppression. In: Proc. of the IEEE Symposium on Research in Security and Privacy (1998)
11. Xu, S., Zhang, W.: PBKM: A secure knowledge management framework (extended abstract). In: NSF/NSA/AFRL Workshop on Secure Knowledge Management, Buffalo, NY (2004)
12. Xu, S., Zhang, W.: Knowledge as service and knowledge breaching. In: IEEE International Conference on Service Computing (SCC 2005) (2005)
13. Kargupta, H., Datta, S., Wang, Q., Sivakumar, K.: On the privacy preserving properties of random data perturbation techniques. In: IEEE International Conference on Data Mining (2003)
14. Xiao, X., Tao, Y.: Personalized privacy preservation. In: ACM SIGMOD International Conference on Management of Data, pp. 229–240 (2006)
15. Evmievski, A., Srikant, R., Agrawal, R., Gehrke, J.: Privacy preserving mining of association rules. In: International Conference on Knowledge Discovery and Data Mining (2002)
16. Rizvi, S.J., Haritsa, J.R.: Maintaining data privacy in association rule mining. In: International Conference on Very Large Data Bases (2002)
17. Merugu, S., Ghosh, J.: Privacy-preserving distributed clustering using generative models. In: IEEE International Conference on Data Mining (2003)
18. Wang, K., Yu, P.S., Chakraborty, S.: Bottom-up generalization: A data mining solution to privacy protection. In: IEEE International Conference on Data Mining (2004)
19. Fung, B.C.M., Wang, K., Yu, P.S.: Top-down specification for informaiton and privacy preservation. In: IEEE International Conference on Data Engineering (2005)
20. LeFevre, K., DeWitt, D.J., Ramakrishnan, R.: Incognito: Efficient fulldomain k-anonymity. In: ACM SIGMOD International Conference on Management of Data (2005)
21. Kifer, D., Gehrke, J.E.: Injecting utility into anonymized datasets. In: ACM SIGMOD International Conference on Management of Data (2006)
22. Wong, R.C.-W., Li, J., Fu, A.W.-C., Wang, K. (α, k)-anonymity: An enhanced k-anonymity model for privacy-preserving data publishing. In: International Conference on Knowledge Discovery and Data Mining (2006)
23. Aggarwal, C., Yu, P.: A condensation approach to privacy preserving data mining. In: International Conference on Extending Database Technology, pp. 183–199 (2004)
24. Xu, L., Krzyzak, A., Suen, C.Y.: Methods of combining multiple classifiers and their applications tohandwriting recognition. IEEE Transactions on Systems, Man and Cybernetics 22(3), 418–435 (1992)

25. Woods, K., Kegelmeyer Jr., W.P., Bowyer, K.: Combination of multiple classifiers using local accuracy estimates. IEEE Transactions on Pattern Analysis and Machine Intelligence 19(4), 405–410 (1997)
26. Oliveira, S., Zaane, O., Saygin, Y.: Secure Association Rule Sharing. In: Dai, H., Srikant, R., Zhang, C. (eds.) PAKDD 2004. LNCS (LNAI), vol. 3056, pp. 74–85. Springer, Heidelberg (2004)
27. Wang, Z., Wang, W., Shi, B.: Blocking inference channels in frequent pattern sharing. In: IEEE International Conference on Data Engineering, pp. 1425–1429 (2007)

Privacy-Preserving Sharing of Horizontally-Distributed Private Data for Constructing Accurate Classifiers

Vincent Yan Fu Tan[1,*] and See-Kiong Ng[2]

[1] Massachusetts Institute of Technology (MIT), Cambridge, MA 02139
vtan@mit.edu
[2] Institute for Infocomm Research (I²R), Singapore 119613
skng@i2r.a-star.edu.sg

Abstract. Data mining tasks such as supervised classification can often benefit from a large training dataset. However, in many application domains, privacy concerns can hinder the construction of an accurate classifier by combining datasets from multiple sites. In this work, we propose a novel privacy-preserving distributed data sanitization algorithm that randomizes the private data at each site independently before the data is pooled to form a classifier at a centralized site. Distance-preserving perturbation approaches have been proposed by other researchers but we show that they can be susceptible to security risks. To enhance security, we require a unique non-distance-preserving approach. We use Kernel Density Estimation (KDE) Resampling, where samples are drawn independently from a distribution that is approximately equal to the original data's distribution. KDE Resampling provides *consistent* density estimates with randomized samples that are *asymptotically independent* of the original samples. This ensures high accuracy, especially when a large number of samples is available, with low privacy loss. We evaluated our approach on five standard datasets in a distributed setting using three different classifiers. The classification errors only deteriorated by 3% (in the worst case) when we used the randomized data instead of the original private data. With a large number of samples, KDE Resampling effectively preserves privacy (due to the asymptotic independence property) and also maintains the necessary data integrity for constructing accurate classifiers (due to consistency).

1 Introduction

Consider the following scenario: A group of hospitals are seeking to construct an accurate *global* classifier to predict new patients' susceptibility to illnesses. It would be useful for these hospitals to pool their data, since data mining tasks such as supervised classification can often benefit from a large training dataset. However, by law, the hospitals cannot release private/sensitive patient

---

* Vincent Tan is supported by the Agency for Science, Technology and Research (A*STAR), Singapore. He performed this work at I²R, A*STAR.

F. Bonchi et al. (Eds.): PinKDD 2007, LNCS 4890, pp. 116–137, 2008.

data (e.g. blood pressure, heart rate, EKG signal, X-ray images). Instead, some form of sanitized data has to be provided to a centralized server for training and classification purposes. It is thus imperative to discover means to protect private information, while at the same time, be able to perform data mining tasks with a masked version of the raw data. Can privacy and accuracy co-exist?

In fact, in many application domains, privacy concerns hinder the combining of datasets generated from multiple sources despite the growing need to share sensitive data. For example, military organizations may now need to share sensitive security information for anti-terrorist operations, financial institutions may need to share private customer data for anti-money laundering operations, and so on. In all these applications, the setting is a Distributed Data Mining (DDM) scenario [21] in which the private data sources are distributed across $L \geq 2$ multiple sites. The L sites each contain private information that should be shared or combined as they are probably inadequate on their own. To protect privacy, the data at each site must undergo randomization locally to give sanitized data for sharing. The sanitized data are pooled as a large training data set to construct an accurate global classifier, as shown in Fig. 1. Note that unlike other previous works [38], in our formulation, there is only a one-way communication to the centralized server required. This further minimizes potential security risks when dealing with large number of sensitive datasets at distributed sites.

In this work, we consider a privacy-preserving distributed data sanitization approach [3] for the purpose of constructing accurate classifiers at the centralized site. Our work is very closely related to privacy-preserving classification [23, 24]. Here we focus on randomizing the data at each site independently

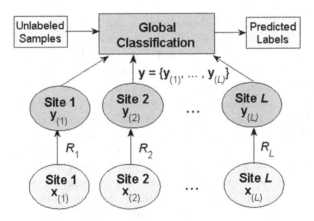

Fig. 1. Privacy-Preserving Distributed Data Mining (DDM) Scenario with $L \geq 2$ sites. $\mathbf{x}_{(l)}$ and $\mathbf{y}_{(l)}$ contain the original and randomized data vectors respectively. The $R_l(\cdot)$'s are the (KDE Resampling) nonlinear randomization operators such that $\mathbf{y}_{(l)} = R_l(\mathbf{x}_{(l)})$. The collection $\mathbf{y} = \{\mathbf{y}_{(l)}\}_{l=1}^{L}$ is to be used as the training data for a *global* classifier. Testing data samples are used for cross-validation of the global classifier at the one centralized site. The non-shaded and shaded cells contain private and randomized data respectively.

before transmitting the data for constructing a global classifier, which is similar to the horizontally partitioned scenario presented in Du et al. [11]. Also, Lindell and Pinkas [23] used Secure Multi-Party (or 2-party) Computation techniques to compute a global decision tree with a (secure) ID3 algorithm. Here, we seek a generic data sanitization approach that can be applied to *any* classification algorithm on numerical data. More recently, Liu et al. [24] and Olivera et al. [28] discussed how random projection-based multiplicative data perturbation can be used for the privacy-preserving DDM scenario. This data perturbation method has several nice properties, including being distance-preserving, which ensures high accuracy in classification and clustering. However, in section 4, we will show that the distance-preserving property can present potential compromises on security of the data. As such, in this work, we will employ a *non-distance-preserving* randomization algorithm for (i) randomizing (sanitizing) the data at the distributed sites (ii) constructing an accurate classifier centrally.

We thus suggest Kernel Density Estimation (KDE) [29, 32, 33] Resampling. KDE Resampling is not new [8] but it has hitherto not been applied to privacy-preserving data mining, to the best of the authors' knowledge. This method possesses some very desirable properties, including *asymptotic independence* and *consistency*, which we will discuss later. Other randomization methods in the literature [1, 2, 7, 24, 28, 38] do not possess these appealing properties. We will exploit these properties to preserve privacy of the distributed data and ensure that the sanitized training samples are still adequate for the construction of accurate classifiers. Note that in our proposed approach, we do not publish the data's distribution/density, since the distribution is fully parameterized by the data records themselves and publishing it would be akin to releasing the private data. Instead, we only transmit the sanitized feature vectors to the centralized site. As shown in Fig. 1, the shaded cells contain the randomized data to be transmitted to the centralized site for the construction of a classifier.

The rest of the paper is structured as follows. In section 2, we discuss in further detail some of the relevant work in sanitization, privacy-preserving DDM and privacy-preserving classification. In section 3, the problem will be formally stated and mathematical notations defined. In section 4, we play the role of a malicious intruder to illustrate the potential security risk in using distance-preserving perturbation methods such as the random projection-based multiplicative data perturbation method [24,28]. KDE Resampling will then be described in section 5. In the same section, we will also discuss the two elegant properties of the samples produced by KDE Resampling. Following that in section 6, we will define two performance metrics and explain the validity. Section 7 details the evaluation experiments and summarizes the main results. Finally, Section 8 concludes our discussion and suggests directions for future work.

2   Related Work

Atallah et al. [3] first considered data sanitization but the work had mainly been applied to association rule mining. Optimal sanitization is NP-hard [3]. We consider classification in this work and a particular randomization method that is computationally tractable.

The addition of randomly generated Independent and Identically Distributed (IID) noise to the original data was then proposed [1,2] for masking the private data. The authors reconstructed the probability density function (PDF) of the data for distribution-based data mining. In addition, they constructed decision-trees based on the noisy data and found that the classification results were similar to that using the original data. Muralidhar et al. [26] comprehensively examined the statistical properties of noise addition.

However, such noise addition has since been shown to be insecure [18,20] and other methods have been proposed. Chen et al. [7] proposed using a rotation-based perturbation technique that ensures low accuracy loss for most classifiers. This perturbation technique was further extended in two papers [24,28] where the authors used a random projection-based multiplicative data perturbation method to perturb the data, while maintaining its utility. These two papers described a randomization method that is distance-preserving. However, it was shown by Caetano [5] that there is data disclosure vulnerability in adopting these distance-preserving approaches. We will further augment Caetano's argument in section 4 by showing that there can be other security risks with distance-preserving approaches. Thus, we will adopt a non-distance-preserving randomization scheme in this paper.

In Zhang et al. [38], an algebraic-based randomization approach was suggested but it involves multiple communication from the server to the sites. This makes it infeasible for extremely large datasets and in scenarios where the communication channels may not be robust (e.g. military scenarios). In our formulation, there is only a one-way communication to the centralized server (Fig. 1). We also do not assume an underlying probability distribution that is parameterized, in contrast to Liew et al. [22]. In addition, we generalize Liew et al. [22] to multiple dependent confidential attributes by using multivariate densities.

Non-randomization approaches have been suggested as well. In Sweeney's papers [31,35], k-anonymization was proposed to generalize databases for preserving privacy. Du et al. [11] approaches the privacy-preserving classification problem from yet another perspective. Using Secure Multi-Party Computation (SMC) techniques [4,34,37], parties can collaborate to deduce a global classifier or regression function or just a general function, like the sum. We will not deal with SMC techniques in this paper as SMC is not as efficient as randomization approaches [30]. However, the obtained results are more accurate than sanitization methods. SMC solutions [4] send and receive input from each of the participating sites thus it is obvious that this method will incur higher communication cost than randomization. For SMC techniques to be collusion-resistant, significant communication is required between the many sites, which make this technique non-practical. Moreover, in SMC the number of participating sites are typically small, which is often not the case in the distributed mining context where number of sites could be few hundreds to several thousands (e.g. surveying, consumer browsing patterns etc.). For a detailed statistical analysis of computation over-head of SMC, the reader is referred to Subramaniam et al. [34].

Yet another method in the literature concerns distributed clustering (unsu-pervised classification) in which the authors chose local models before combining them to give a global model via optimization of information theoretic quanti-ties [25]. We focus on supervised classification here, but our method can be extended for clustering applications.

As mentioned, we will be adopting a technique known as KDE Resampling [8]. This method has many appealing properties, including *asymptotic independence* and *consistency*, which will be fully explained in section 5. Besides these two appealing properties, Indyk and Woodruff [19] also demonstrated that sampling achieves perfect privacy in 2-party polylog-communication $\mathcal{L}_2$ distance approx-imations. Motivated by promising nature of sampling, we explore its proper-ties when applied to a distributed scenario and the subsequent construction of classifiers.

In terms of evaluation metrics, privacy has typically been measured using *mutual information* [1] as well as *privacy breaches* [14]. Mutual information is an average measure of disclosure while privacy breaches examine the worst-case scenario. Because of our Distributed Data Mining (DDM) setting, we will measure privacy in this paper using a new metric – the Distributed Aggregate Privacy Loss, $\mathcal{DAPL}$, which is related to the mutual information. Our measure is advantageous because it explicitly takes into account the *distributed* nature of the data mining scenario. Moreover, the privacy breach measure was primarily used in the context of association rule mining [15] while in this paper, we are concerned with supervised classification using the sanitized data from the L independent data sites.

3   Problem Definition and Notation

We represent the storage of private information in the form of d-dimensional real row vectors $\mathbf{x}_1, \ldots, \mathbf{x}_N$, where N is the number of individuals (subjects) and d is the number of attributes. These row vectors can be vertically concatenated into a $N \times d$ matrix $\mathbf{x}$ such that

$$\mathbf{x} \triangleq \begin{bmatrix} \mathbf{x}_1 \\ \vdots \\ \mathbf{x}_N \end{bmatrix} = \begin{bmatrix} x_{11} & \cdots & x_{1d} \\ \vdots & \ddots & \vdots \\ x_{N1} & \cdots & x_{Nd} \end{bmatrix}. \tag{1}$$

These N individuals are associated with N targets (class labels) $t_1, \ldots t_N$. The class labels are typically not regarded as sensitive/private data [2, 38] and thus they do not have to be randomized.

We then assume that there are L (for $2 \leq L \leq N$) distributed data sites (private) and 1 centralized (untrusted) server (Fig. 1), where the sanitized data are sent to for constructing an accurate classifier using the combined training data. Each data site possesses the private information of N_l individuals, with

$\sum_{l=1}^{L} N_l = N$. As in Fig. 1, we use the notation $\mathbf{x}_{(l)}$ for the $N_l \times d$ matrix that contains the N_l data vectors at site l. Thus,

$$\mathbf{x}_{(l)} \triangleq \left[\mathbf{x}_{(l,1)}^T \cdots \mathbf{x}_{(l,N_l)}^T \right]^T, \quad 1 \le l \le L, \tag{2}$$

where $\mathbf{x}_{(l,j)}$ for $1 \le j \le N_l$ is a sample vector at site l. Thus, $\mathbf{x}$ can alternatively be written as

$$\mathbf{x} = \left[\mathbf{x}_{(1)}^T \cdots \mathbf{x}_{(L)}^T \right]^T. \tag{3}$$

Furthermore, we assume that the row vectors in $\mathbf{x}_{(l)}$ are drawn from IID random vectors with PDF $f_{\mathbf{X}_{(l)}}\left(\mathbf{x}_{(l)}\right)$. We seek to find a randomization scheme for site l, R_l such that

$$\mathbf{y}_{(l)} = R_l(\mathbf{x}_{(l)}), \quad 1 \le l \le L. \tag{4}$$

and $R_l : \mathbb{R}^{N_l \times d} \to \mathbb{R}^{M_l \times d}$ is the nonlinear randomization operator that maps N_l row vectors in $\mathbf{x}_{(l)}$ to M_l row vectors in $\mathbf{y}_{(l)}$. $\mathbf{y} = \{\mathbf{y}_{(l)}\}_{l=1}^L$ is then sent to the centralized server, along with the N associated targets $t_1, \ldots, t_N$, where classification can then be done using randomized data as training samples[1]. The centralized server will use the pooled randomized/sanitized data as training samples to build a classifier. We will show that the classification results using these randomized data as training samples are compatible to the classification results using the original private data as training samples. Before that, let us first examine why distance-preserving approaches may be vulnerable to attacks by malicious intruders.

4   Risk of Distance-Preserving Randomization

In this section, we play the role of a malicious attacker and attempt to deduce information such as the bounds on private data sanitized with a distance-preserving perturbation method such as the random projection-based multiplication method [7,24,28]. Caetano [5] had showed previously that the randomized data can be vulnerable to disclosure. We will further augment his argument with two lemmas here.

Lemma 1. *Assume a Distributed Data Mining (DDM) scenario with $L = 2$ sites which contain private data matrices $\mathbf{x}_{(1)}$ and $\mathbf{x}_{(2)}$ respectively. Upon randomization using the random projection-based multiplicative data perturbation method[2], we get $\mathbf{y}_{(1)} = \mathbf{R}\mathbf{x}_{(1)}$ and $\mathbf{y}_{(2)} = \mathbf{R}\mathbf{x}_{(2)}$ respectively. Let the matrix $\mathbf{x}_{(1)}$ have the structure as follows:*

$$\mathbf{x}_{(1)} \triangleq \left[\tilde{\mathbf{x}}_{(1,1)} \cdots \tilde{\mathbf{x}}_{(1,d)} \right] \tag{5}$$

---

[1] Note that random vectors are denoted in boldface upper case and the realization is denoted is boldface lower case. For e.g. , $\mathbf{X}_{(l)}$ is a random vector and its realization is $\mathbf{x}_{(l)}$.

[2] In [24], $\mathbf{R} \in \mathbb{R}^{K \times N}$, a random matrix was used to perturb the data via a linear transformation to a lower-dimensional subspace i.e. $K < N$.

and its columns $\tilde{\mathbf{x}}_{(1,i_1)}$ to be defined as

$$\tilde{\mathbf{x}}_{(1,i_1)} = \left[\tilde{x}_{(1,i_1,1)} \cdots \tilde{x}_{(1,i_1,N_1)} \right]^T, \quad 1 \le i_1 \le d \tag{6}$$

Let the other matrices $\mathbf{x}_{(2)}$, $\mathbf{y}_{(1)}$ and $\mathbf{y}_{(2)}$ have similar structures. Further, suppose we have $\|\widehat{\tilde{\mathbf{x}}_{(2,i_2)}}\|$, an estimate of the norm[3] of the i_2^{th} column of $\mathbf{x}_{(2)}$ for any $1 \le i_2 \le d$, then

$$\|\tilde{\mathbf{x}}_{(1,i_1)}\| \ge \gamma_1, \tag{7}$$

for all $1 \le i_1 \le d$, where $\gamma_1 > 0$ is a constant.

All proofs can be found in the Appendix. Lemma 1, gives us a lower bound for the norm of *all* the columns of the matrix $\mathbf{x}_{(1)}$, given an estimate of *just one* column of the matrix $\mathbf{x}_{(2)}$. Clearly, there is an obvious security risk, especially if the private values are susceptible to being leaked. Lemma 2 builds on this to infer a lower bound on any private data value given other data values.

Lemma 2. *Assume exactly the same DDM scenario as in Lemma 1 and that we have estimates for all the elements of $\tilde{\mathbf{x}}_{(1,i_1)}$ except the qth element $\tilde{x}_{(1,i_1,q)}$ i.e. we are given the set*

$$\mathcal{A}_{i_1,\backslash q} = \{\tilde{x}_{(1,i_1,k)} | \tilde{x}_{(1,i_1,k)} \in \tilde{\mathbf{x}}_{(1,i_1)}, k \ne q\}. \tag{8}$$

Then,

$$\left| \tilde{x}_{(1,i_1,q)} \right| \ge \gamma_2, \tag{9}$$

for all $1 \le q \le N_1$, where $\gamma_2 > 0$ is a constant.

Lemma 2 shows that if a malicious attacker were to obtain estimates of data values except the qth element for the data vectors in any of the d dimensions, he or she will be able to infer lower bounds on the private data value he does not possess i.e. $|\tilde{x}_{(1,i_1,q)}|$. This is a potential security breach. Intuitively, there is such a breach because along with the preservation of distances, the 'ordering' of the samples is also preserved. This reasoning (and lemmas) can be extended to the case where $L > 2$. Together with Caetano's argument [5], there is clearly a need for a new randomization method for privacy-preserving classification that is not distance-preserving. In light of the limitations of the additive method [1,2,26] and the random projection perturbation method [7,24,28], in this work, we will use KDE Resampling, which is a non-distance-preserving randomization algorithm, for data sanitization.

5   KDE Resampling for Data Sanitization

In this section, we will detail KDE Resampling and discuss some elegant and useful properties of the randomized samples. We will also comment on its computational tractability and compare it to the more inefficient SMC methods [23].

---

[3] Any valid l_p (for $p \ge 1$) norm can be used.

5.1 Resampling from Reconstructed PDF

For each of the L data sites (refer to Fig. 1), we will generate M_l independent vectors in $\mathbf{y}_{(l)}$ with approximately the same density as the original N_l vectors in $\mathbf{x}_{(l)}$. M_l and N_l do not necessarily have to be equal. The algorithm takes place in two steps. Firstly, we will approximate the PDF of the vector in $\mathbf{x}_{(l)}$ using Parzen-Windows Estimation [29] also known as KDE [10, 32, 33]. Then we will sample M_l vectors from this PDF, which we denote $\mathbf{y}_{(l)}$.

Kernel Density Estimation. As mentioned, for data site l, we will construct the multivariate PDF using the N_l vectors in the $N_l \times d$ matrix $\mathbf{x}_{(l)}$, which we denote $\mathbf{x}_{(l,1)}, \ldots, \mathbf{x}_{(l,N_l)}$. This is given by

$$\hat{f}_{\mathbf{X}_{(l)}} \left(\mathbf{x}_{(l)}; \mathbf{x}_{(l,1)}, \ldots, \mathbf{x}_{(l,N_l)} \right) = \frac{1}{N_l} \sum_{j=1}^{N_l} K \left(\mathbf{x}_{(l)} - \mathbf{x}_{(l,j)}; \mathbf{h}_l \right), \tag{10}$$

where $K \left(\mathbf{x}_{(l)} - \mathbf{x}_{(l,j)}; \mathbf{h}_l \right)$ is the Epanechnikov[4] (a truncated quadratic) kernel parameterized by $\mathbf{h}_l$, the vector of bandwidths. In one dimension, K is given by

$$K_1(x; h) = c\,h^{-1} \left(1 - \left(\frac{x}{h} \right)^2 \right) \mathbb{I}\{|x| \le h\}, \tag{11}$$

where c is the normalizing constant. The multivariate version of the Epanechnikov kernel is a straightforward generalization by taking products of the univariate kernel in Eq (11). $K\left(\cdot; \mathbf{h}_l\right)$, a scalar kernel function, has to satisfy the following properties for Eq (10) to be a valid PDF [16].

$$K\left(\mathbf{x}; \mathbf{h}_l\right) \ge 0, \ \forall \mathbf{x} \in \mathbb{R}^d, \quad \int_{\mathbb{R}^d} K\left(\boldsymbol{\xi}; \mathbf{h}_l\right) d\boldsymbol{\xi} = 1. \tag{12}$$

Example 1. An illustration of how the univariate KDE works for $N = 7$ is shown in Figure 2. The kernels are centered on the realizations of the multi-modal random variable and the sum is an approximation to the true PDF. Notice that, consistent with intuition, more probability mass is placed in areas where there are more realizations of the random variable.

The selection of the bandwidth vector $\mathbf{h}_l \in \mathbb{R}^d$ is a very important consideration in any KDE and will be discussed in section 5.2. For optimal performance and accuracy of the KDE, $\mathbf{h}_l$ is to be a function of the number of samples N_l. We note that the Kernel Density Estimate in Eq (10) is a function of $\mathbf{x}_{(l)}$ and it is parameterized by the realizations of IID random vectors $\mathbf{x}_{(l,1)}, \ldots, \mathbf{x}_{(l,N_l)}$ present at site l. Thus, the distribution cannot be published. Instead, we will transmit the M_l randomized data vectors from site l to the centralized site for the construction of a classifier.

---

[4] The Epanechnikov kernel is optimal in the l_2 sense [8].

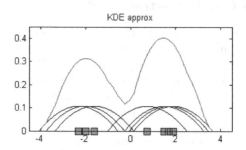

Fig. 2. Illustration of KDE approximation for estimation of the multimodal PDF. The *boxes* show the $N = 7$ independent realizations of the multimodal random variable. The individual Epanechnikov kernels ($h = 1.75$) are centered at the realizations. Their *sum*, as detailed in Eq (10), is the Kernel Density Estimate (KDE), which is the sum of the Epanechnikov kernels.

Remark 1. For the sake of convenience, we chose M_l and N_l to be equal. However, in practice, they do not have to be equal. In fact, one can sample fewer data vectors than N_l, for example to choose $M_l = N_l/2$. From our experiments, the classification results do not change significantly when $M_l = N_l/2$. We refer the reader to Devroye's book [9, Chapter 14] for a more rigorous treatment on the selection of M_l.

We will subsequently abbreviate the estimate of the true PDF by $\hat{f}_l \triangleq \hat{f}_{\mathbf{X}_{(l)}}\left(\mathbf{x}_{(l)}; \mathbf{x}_{(l,1)}, \ldots, \mathbf{x}_{(l,N_l)}\right)$ and the true PDF by $f_l \triangleq f_{\mathbf{X}_{(l)}}\left(\mathbf{x}_{(l)}\right)$.

Resampling. Equipped with the non-parametric estimate of the true PDF $\hat{f}_l$, we will then sample from this PDF to obtain M_l independent samples $\mathbf{y}_{(l,1)}, \cdots, \mathbf{y}_{(l,M_l)}$. Noting that the random vector $\mathbf{X}_{(l)} = (1/N_l) \sum_{j=1}^{N_l} \mathbf{X}_{(l,j)}$ is a mixture density – it does not have to be constructed explicitly before random samples are taken. Instead we will sample for a random integer r from 1 to N_l. Following that we will sample a random vector from the rth kernel $K\left(\mathbf{x}_{(l)} - \mathbf{x}_{(l,r)}; \mathbf{h}_l\right)$. The resampling algorithm is summarized in Algorithm 1.

5.2   Discussion

In any privacy-preserving data mining research, the two key questions are: Has privacy been preserved? Can the randomized vectors be used for data mining purposes? In this section, we will state some very important and salient results from [10]. These results show that the randomized samples $\mathbf{y}_{(l)}$ at each of the L sites are *asymptotically independent* of the original samples $\mathbf{x}_{(l)}$ at the respective L sites. Also, the KDE is *consistent*. We will explain why these two properties are desirable in subsequent sections. We will explain that privacy can indeed be preserved while the randomized samples can be employed for data mining.

KDE Resampling Algorithm

Data : $\mathbf{x}_{(l,1)}, \ldots, \mathbf{x}_{(l,N_l)}$ for all $1 \leq l \leq L$

Result : $\mathbf{y}_{(l,1)}, \ldots, \mathbf{y}_{(l,M_l)}$ for all $1 \leq l \leq L$

for $l \leftarrow 1$ to L do

    for $i \leftarrow 1$ to d do

        $\hat{\sigma}_{l,i}$ = Standard deviation in dimension i;

        $h_{l,i}$ = Bandwidth in dimension i (c.f Eq (16));

    endFor

    for $j \leftarrow 1$ to M_l do

        r = Random integer from 1 to N_l inclusive;

        $\mathbf{y}_{(l,j)}$ = Random sample vector from r^{th} Epanechnikov kernel $K\left(\mathbf{y}_{(l,j)} - \mathbf{x}_{(l,r)}; \mathbf{h}_l\right)$;

    endFor

endFor

Algorithm 1. KDE Resampling

Asymptotic Independence. Asymptotic independence implies that the randomized samples are independent of the original samples as the number of samples N_l tends to infinity. If the joint density of $\mathbf{X}_{(l)}$ and $\mathbf{Y}_{(l)}$ is denoted as $f_{\mathbf{X}_{(l)}, \mathbf{Y}_{(l)}}\left(\mathbf{x}_{(l)}, \mathbf{y}_{(l)}\right)$ and the marginals as $f_{\mathbf{X}_{(l)}}\left(\mathbf{x}_{(l)}\right)$ and $f_{\mathbf{Y}_{(l)}}\left(\mathbf{y}_{(l)}\right)$, then asymptotic independence can be expressed mathematically as

$$\limsup_{N_l \to \infty} |\Delta_{N_l}| = 0, \tag{13}$$

where the difference between the joint and product of the marginals is defined as

$$\Delta_{N_l} \triangleq f_{\mathbf{X}_{(l)}, \mathbf{Y}_{(l)}}\left(\mathbf{x}_{(l)}, \mathbf{y}_{(l)}\right) - f_{\mathbf{X}_{(l)}}\left(\mathbf{x}_{(l)}\right) f_{\mathbf{Y}_{(l)}}\left(\mathbf{y}_{(l)}\right), \tag{14}$$

and the supremum in Eq (13) is over all possible realizations of $\mathbf{x}_{(l)}$ and $\mathbf{y}_{(l)}$.

Another important point is that asymptotic independence is dependent on how we select the bandwidth vector $\mathbf{h}_l$ in Eq (10). If $\mathbf{h}_l$, a function of N_l, satisfies

$$h_{l,i} \xrightarrow{P} 0, \quad \text{and} \quad N_l h_{l,i}^d \xrightarrow{P} \infty, \tag{15}$$

as $N_l \to \infty$ then *asymptotic independence* will be achieved [10]. Note that $h_{l,i}$ is the ith element of the bandwidth vector $\mathbf{h}_l$. In our experiments, we are going to use the Scott's 'rule-of-thumb' [32] to select $\mathbf{h}_l$. Thus,

$$h_{l,i} = \left(\frac{4}{d+2}\right)^{1/(d+4)} N_l^{-1/(d+4)} \hat{\sigma}_{l,i}, \tag{16}$$

where $\hat{\sigma}_{l,i}$ is the unbiased estimate of the standard deviation in the ith dimension at the lth site. Scott's 'rule-of-thumb' satisfies both the asymptotic conditions and thus, we have asymptotically independent samples. Since the samples are asymptotically independent, probabilistic inference cannot be performed based on the randomized samples $\mathbf{y}_{(l)}$ if N_l is sufficiently large. This is very often the case in practical data mining scenarios, where datasets are extremely large. Privacy will thus be preserved.

Another way to illustrate this is using the privacy loss measure based on mutual information [1]. Indeed, if N_l is sufficiently large (like in most practical data mining applications), the mutual information $I(\mathbf{x}_{(l)}; \mathbf{y}_{(l)})$ will be close to zero (because of asymptotic independence) and thus, the privacy loss $\mathcal{P}\left(\mathbf{x}_{(l)}; \mathbf{y}_{(l)}\right) = 1 - 2^{-I(\mathbf{x}_{(l)}; \mathbf{y}_{(l)})}$ will also be low. In section 6, we will define a new privacy metric, $\mathcal{DAPL}$, and argue that the asymptotic independence of the randomized samples will result in low privacy loss when a large number of samples are available. This property ensures that KDE Resampling is especially effective for preserving the privacy of large datasets i.e. large N_l's.

Consistency of KDE. It is well known [10,33] that the KDE $\hat{f}_l$, as defined in Eq (10) is consistent i.e.

$$\lim_{N_l \to \infty} \mathbb{E}\left[\int \left|\hat{f}_l - f_l\right|\right] = 0, \quad 1 \le l \le L, \tag{17}$$

if the asymptotic conditions in Eq (15) are satisfied. This means that as the number of samples at each site N_l becomes large, the KDE $\hat{f}_l(\cdot)$ becomes increasingly accurate. This property is important and useful because we can treat the collection of randomized samples at all the L sites $\{\mathbf{y}_{(l)}\}_{l=1}^{L}$ as the training data for supervised classification purposes since the distribution it is drawn from is consistent.

Remark 2. We note that because of resampling, our randomization algorithm does not suffer from the problems of [7,24,28] that were highlighted in section 4 – namely that of being able to derive bounds on private data given other (relevant) private information. This is one of the key advantages of our novel randomization technique as it removes the inherent ordering of the feature vectors by resampling randomly.

Low Computational Complexity. The random vector $\mathbf{X}_{(l)} = (1/N_l) \sum_{j=1}^{N_l} \mathbf{X}_{(l,j)}$ is a mixture density with N_l components. Thus, we do not need to construct the full KDE. This is typically the bottleneck for any algorithm that uses the Kernel Density Estimate (KDE). Thus the randomized vectors can be obtained simply by:

1. First, estimating the kernel bandwidths $h_{l,i}$, $\forall (l, i) \in \{1, \ldots, L\} \times \{1, \ldots, d\}$ using Eq (16).
2. Generating a random (integer) index r from 1 to N_l.
3. Then drawing a random sample vector from the rth multivariate Epanechnikov kernel.

This is detailed more precisely in Algorithm 1. Each step in the algorithm is tractable. There is no multi-way communication between the parties, unlike in SMC [34, 37]. In conclusion, the KDE Resampling algorithm is computationally feasible.

Possible Application to Horizontally or Vertically Partitioned Data. We have presented a randomization algorithm for the purpose of randomizing *horizontally partitioned* data over L sites. The extension to the *vertically partitioned* scenario, where different sites hold different attributes, is not trivial unless attributes are assumed to be independent as in [36]. KDE Resampling requires multiple full data vectors to be most effective and accurate.

6   Performance Metrics

For evaluation, the two performance metrics that we will use to quantify privacy and accuracy are the *Distributed Aggregate Privacy Loss* $\mathcal{DAPL}$ and the *Deterioration of Classification* ϕ respectively.

6.1   Distributed Aggregate Privacy Loss $\mathcal{DAPL}$

The privacy loss is a function of *mutual information* [1], which depends on the *degree of independence* between the randomized samples and the original samples. We have decided to design our privacy metric based on mutual information because the task we are handling is supervised classification. In Evfimievski [14], the notion of security breach was raised. In this work, we focus more on privacy loss, which is an average measure of privacy disclosure. Moreover, the privacy measures proposed by the same paper were more applicable to association rule mining [15]. Thus, in this paper, we use $\mathcal{DAPL}$, which is intimately related to mutual information. Mutual information measures the average amount of information disclosed when the randomized data is revealed. Indeed, [8] also mentioned that

> "... for the sake of asymptotic sample independence, it suffices that the expected l_1 distance between $[\hat{f}_l]$ and $[f_l]$ tends to zero with $[N_l]$."

Because expected l_1 distances provide us with the degree of independence, we will define our privacy loss as a weighted average of expected l_1 distances.

Definition 1. *The* Distributed Aggregate Privacy Loss $\mathcal{DAPL}$ *is defined as half of the weighted average of the expected l_1 distance between the estimate $\hat{f}_l$ and f_l, over the L sites. That is,*

$$\mathcal{DAPL} \triangleq \frac{1}{2} \left(\sum_{l=1}^{L} c_l \, \mathbb{E} \left[\int \left| \hat{f}_l - f_l \right| \right] \right), \tag{18}$$

where $c_l \triangleq N_l/N$ for $1 \leq l \leq L$ is the proportion of samples at site l. Clearly, $0 \leq \mathcal{DAPL} \leq 1$.

The $\mathcal{DAPL}$ is low (≈ 0) when privacy loss is low and vice versa. Finally, we emphasize that our privacy loss metric $\mathcal{DAPL}$ is related to, but not exactly identical to, the privacy loss metric defined in [1]. The difference is in our considering of the distributed scenario here. Furthermore, we also measure the degree of independence using the expected l_1 distance between $\hat{f}_l$ and f_l as opposed to using mutual information. Since the l_1 distance $\to 0$ as $N_l \to \infty$ [33], the asymptotic independence property ensures low $\mathcal{DAPL}$ when N_l is large.

We emphasize that Eq (18) is a reasonable *privacy measure* because independence of the original and randomized data samples is measured in terms of expected l_1 distances between the original density and the KDE [8].

Example 2. For an *intuitive* feel of Eq (18), let us consider a single site with N_l samples. Suppose N_l is large, then an accurate KDE will be constructed. Subsequently, because we sample from a randomly chosen kernel (out of the N_l kernels whose means are the original data vectors), the resulting randomized data vectors will be approximately independent of the original samples. On the other hand, suppose N_l is small, say only two samples, then the resulting randomized data vectors will be strongly dependent on the positions, in $\mathbb{R}^d$, of the two original samples, resulting in less 'randomness' and greater privacy loss. Another relevant paper by Dwork [13] applies the definition of *differential privacy* to the case of distributed computations, much like our paper.

6.2   Deterioration of Classification ϕ

In any supervised classification algorithm, the usual performance metric is the probability of error, which is also known as the classification error and is defined as [17]

$$P(err) \stackrel{\triangle}{=} 1 - \sum_{i=1}^{|\mathcal{C}|} \int_{\Omega_i} p\left(\boldsymbol{\xi}|\mathcal{C}_i\right) P(\mathcal{C}_i)\, d\boldsymbol{\xi}, \tag{19}$$

where $P(\mathcal{C}_i)$ is the prior probability (known *a priori*) of class $\mathcal{C}_i$ and $\int_{\Omega_i} p\left(\boldsymbol{\xi}|\mathcal{C}_i\right) P(\mathcal{C}_i)\, d\boldsymbol{\xi}$ is the conditional probability of correct classification[5] for given the sample is from $\mathcal{C}_i$. $|\mathcal{C}|$ denotes the total number of classes.

Definition 2. *The Deterioration of Classification ϕ is defined as:*

$$\phi \stackrel{\triangle}{=} P_{rand}(err) - P_{ori}(err), \tag{20}$$

where $P_{ori}(err)$ (resp. $P_{rand}(err)$) is the classification error using the original (randomized) samples as training data.

Clearly, the closer ϕ is to zero, the greater the utility of the randomized samples and the higher the accuracy. So, we want ϕ to be as small as possible.

---

[5] Also known as a 'hit' in the detection theory literature.

7   Simulation Results

In this section, we will detail the classification experiments on five diverse datasets with continuous, numerical values using three different classifiers. We will consider the distributed setting in Fig. 1. We will empirically show that the classification error is invariant to original and randomized data being used as training examples.

7.1   Experimental Setup of Distributed Setting

In all our experiments, we consider a *distributed* scenario (like in Fig. 1) with L sites, where L is taken over a range of integers. Hence, suppose there are N data points (and N is a multiple of L), then each site will contain $N_l = N/L$ points. If N is not a multiple of L, minor adjustments are made. Each of the N_l data points at the L sites are randomized using the algorithm detailed section 5. The data is then pooled to the centralized site for the construction of various classifiers. The classification accuracy is compared to the baseline – the result when the data is not randomized.

Table 1. Our five datasets from LIBSVM and the UCI Machine Learning Repository

Dataset	#Class	#Dim(d)	#Trg(N)	#Test
Iris	2	4	120	30
SVMGuide1	2	4	3089	4000
Diabetes	2	8	576	192
Breast-Cancer	2	10	512	171
Ionosphere	2	34	263	88

7.2   Datasets

We obtained five numerical datasets from LIBSVM [6] and the UCI Machine Learning Repository [27]. These are summarized in Table 1. These include the common Iris Dataset and the more difficult Pima Indians Diabetes Dataset.

We pre-process the data by normalizing the values in each dimension to the interval $[-1, 1]$ before the randomization and classification processes. As mentioned in the above section, for each dataset, we performed the randomization followed by classification using a different number of sites L. For example, for the Iris dataset[6] (see Figure 3), L was chosen to be from 1 to 4.

For the purpose of validating the classification accuracy, the raw data (except for the SVMGuide1 dataset[7]) was randomly split into a training set ($\approx 75\%$)

---

[6] For the Iris dataset, we merged the Setosa and the Versicolour classes into one single class so that we have a binary classification problem. Even though all of our analysis can be extended to the multi-class scenario, it seems not too relevant for the questions addressed here.

[7] SVMGuide1 had already been partitioned into training and testing data *a priori* [6] and thus we use the given partitioning to test the classifiers constructed.

and a testing set ($\approx 25\%$). This is also commonly-known in the literature as random subsampling *4-fold cross-validation*. For consistent results, we averaged the classification errors over 100 independent random seeds.

Finally, the number of vectors we resampled M_l is the same as the number of original vectors N_l at all L sites. However, as argued in section 5.1, M_l does not have to be the same as N_l. For brevity, we only report the case where $M_l = N_l$ in this section. However, an obvious advantage of using a fewer number of samples is reduction in complexity.

7.3   Classification Techniques

We used three standard classification techniques on the combined randomized data from the L sites $\mathbf{y}_{(1)}, \ldots, \mathbf{y}_{(L)}$ and the original data in $\mathbf{x}$. These techniques include:

1. Artificial Neural Networks (ANN) by trained by error backpropagation.
2. k-Nearest Neighbors classifier (kNN) with $k = 11$.
3. Naïve Bayes classifier (NB) with each attribute or dimension (d) assumed to follow a Gaussian distribution.

The details of these classification techniques can be found in any standard pattern classification text. See for instance [17, 12].

First, we used the above three classification techniques to obtain initial classification results on the original data samples. These are shown in Table 2. These results, denoted $P_{ori}(err)$, will be compared to $P_{rand}(err)$, the classification results on the randomized samples in a distributed setting with L sites. The basis for comparison is their difference $\phi \overset{\triangle}{=} P_{rand}(err) - P_{ori}(err)$ (cf. section 6.2).

Table 2. $P_{ori}(err)$ using original data $\mathbf{x}$ as training samples for various classification techniques. Refer to Figs. 3 to 7 for $\phi = P_{rand}(err) - P_{ori}(err)$.

Dataset	ANN	kNN	NB
Iris	0.0352	0.0346	0.1230
SVMGuide1	0.0389	0.0328	0.0695
Diabetes	0.2441	0.2611	0.2263
Breast-Cancer	0.0312	0.0214	0.0396
Ionosphere	0.1255	0.1522	0.0000

7.4   Results of Distributed Experiments

The results are shown in Figures 3 to 7. Sub-figures (a) show the values of the Deterioration of Classification ϕ, which is defined in Eq (20). The three lines show ϕ for different classification techniques ANN (in crosses – x), kNN (in circles – o) and NB (in plusses – +). We plot the Distributed Aggregate Privacy Loss $\mathcal{DAPL}$ for the two classes in sub-figures (b) (in crosses – x and circles – o). From the plots, we made the following observations.

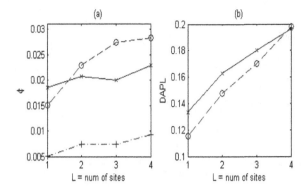

Fig. 3. Iris; (a) Deterioration of Classification ϕ (Key: x - ANN, o - kNN, + - NB); (b) Distributed Aggregate Privacy Loss $\mathcal{DAPL}$ (Key: x - Class $\mathcal{C}_1$, o - Class $\mathcal{C}_2$); L was chosen to be from 1 to 4 for the commonly-encountered Iris dataset. Note from (a) that ϕ increases as L is increased, as the number of data records N_l at each site is reduced. However, the Deterioration of Classification ϕ is less than 3% and in this case, the Naïve Bayes classifier (+) performs the best (least deterioration). $\mathcal{DAPL}$ increases as the number of sites L increases because there are fewer samples at each site (cf. example in section 6.1).

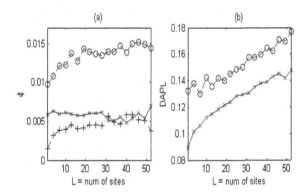

Fig. 4. SVMGuide1; (a) Deterioration of Classification ϕ (Key: x - ANN, o - kNN, + - NB); (b) Distributed Aggregate Privacy Loss $\mathcal{DAPL}$ (Key: x - Class $\mathcal{C}_1$, o - Class $\mathcal{C}_2$); The SVMGuide1 dataset describes an astroparticle application. For this dataset, we observe that the ϕ does not increase significantly across L. There is little correlation between the number of sites L and the classification errors $P_{rand}(err)$ or the Deterioration of Classification ϕ. Comparing the results in (a) with Table 2, we observe that the deterioration is not too severe. However, as expected, Privacy Loss $\mathcal{DAPL}$ increases as the number of sites L increases for the same reason as stated in the caption for the Iris dataset.

The classification errors $P_{rand}(err)$ and $P_{ori}(err)$ are close. This can be seen from sub-figures (b) for each of the five datasets, where ϕ deviates from zero by at most only 3%. In general, Naïve Bayes (NB) and Artificial Neural Networks

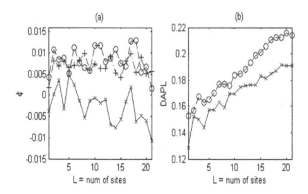

Fig. 5. <u>Diabetes</u>; (a) Deterioration of Classification ϕ (Key: x - ANN, o - kNN, + - NB); (b) Distributed Aggregate Privacy Loss $\mathcal{DAPL}$ (Key: x - Class $\mathcal{C}_1$, o - Class $\mathcal{C}_2$); The diabetes dataset exhibits the same characteristics as the SVMGuide1 dataset. However, in this case, it is somewhat surprising to note that in most cases, the ANN classifier constructed based on the randomized samples results in a lower classification error as compared to the one constructed based on the original samples.

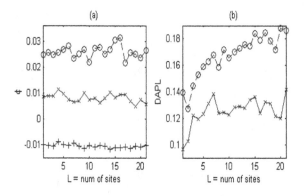

Fig. 6. <u>Breast-Cancer</u>; (a) Deterioration of Classification ϕ (Key: x - ANN, o - kNN, + - NB); (b) Distributed Aggregate Privacy Loss $\mathcal{DAPL}$ (Key: x - Class $\mathcal{C}_1$, o - Class $\mathcal{C}_2$); For the Breast-Cancer dataset, the Deterioration of Classification ϕ stays fairly constant across all L. Indeed, the NB classifier (+) constructed based on the randomized samples is better (improves by 1%) than the classifier constructed using the original samples. Again, we observe that, as expected, Privacy Loss $\mathcal{DAPL}$ increases as L increases because the number of samples at each site N_l decreases.

(ANN) perform better as compared to k-Nearest Neighbors (kNN) as the Deterioration of Classification ϕ is closest to zero for all the datasets for NB and ANN.

Except for the Iris dataset, there is no correlation between the number of sites L and the classification errors $P_{rand}(err)$ or the Deterioration of Classification ϕ. Consequently, the randomized data is still amenable to data mining tasks

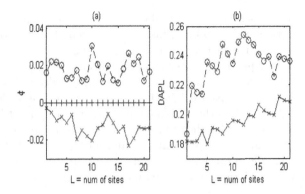

Fig. 7. Ionosphere; (a) Deterioration of Classification ϕ (Key: x - ANN, o - kNN, + - NB); (b) Distributed Aggregate Privacy Loss $\mathcal{DAPL}$ (Key: x - Class $\mathcal{C}_1$, o - Class $\mathcal{C}_2$); In this dataset, we observe that the ANN classifier constructed based on the pooled randomized data samples from L sites results in a smaller classification error. The kNN classification technique does results in a worse classification error but, in the worst case, the performance does not deteriorate by more than 3%. Compared with the baseline in Table 2, we see that this deterioration is not too severe.

irregardless of L. This is because of the consistency of the KDE as discussed in section 5.2.

Finally, from sub-figures (b), we observe that there is a general increasing trend for the $\mathcal{DAPL}$. This is because as the number of sites L increases, the number of individuals at each site N_l decreases. Consequently, the expected l_1 difference between the the original and reconstructed PDFs increases and the privacy loss $\mathcal{DAPL}$ also increases. Thus, with the use of KDE Resampling, there is a *compromise* between L and $\mathcal{DAPL}$. The $\mathcal{DAPL}$ can be further reduced by improving the simple sampling algorithm suggested in Algorithm 1. This can be done by improving selection of the optimal of $\mathbf{h}_l$ [32, 33] by possibly optimizing over non-diagonal bandwidth matrices, which enhances generality but increases complexity.

8   Conclusions

In this paper, we have suggested a novel method for data *sanitization* for the purpose of sharing private data at distributed data sites for constructing a classifier. In our setup, we are provided with N_l data records at each site and we apply the randomization algorithm at each site independent of other sites. Then the data is pooled together for classification at the centralized site. As mentioned in the introduction, this problem has ramifications in a variety of settings, including the sharing of patients' private records and for collaborations across military or financial organizations for various security operations.

We employed Kernel Density Estimation (KDE) Resampling to sample for new, representative data vectors given the original private data. Our experiments on five datasets conducted in a distributed data setting illustrate that resampling provides sanitized/randomized data samples that can be adequately employed for a particular data mining task – supervised classification. In summary, our data sanitization algorithm has the following advantages over some existing approaches for privacy-preserving classification.

KDE Resampling provides samples that are *asymptotically independent* and the KDE is *consistent*. We have explained, that the former ensures low privacy loss, while the latter preserves the data's integrity and hence, its utility.

We have shown that the classification errors using the *randomized data* collated from the distributed sites as training samples differs from that using the *original data* as training samples by less than 3% for all the datasets. We have also shown that various data mining algorithms can be applied on the randomized training data.

In contrast to random projection-based multiplicative data perturbation methods [7, 24, 28], a malicious intruder cannot establish bounds on the private data using KDE Resampling. In fact, Caetano [5] also argued that random projection-based randomization may be susceptible to disclosure. KDE Resampling thus ensures greater security as it involves an element of random *swapping*, which enhances privacy by losing the ordering of the feature vectors.

In contrast to [38], our framework as shown in Figure 1 does not involve multi-way communication from the centralized server to the individual sites and vice versa. Since our technique involves only a *one-way* communication from the sites to the server, it is feasible for large datasets. Besides, one-way communication reduces the risk of inadvertent disclosure of private data.

Although SMC techniques may provide better privacy protection and accuracy as compared to randomization methods, they suffer from inefficiency [23, 30, 34]. Our algorithm generates samples in an efficient fashion, because each step of the algorithm is tractable and there is no need for multi-way communication.

Finally, we hope to re-examine the issue of the privacy metric. Since $\mathcal{DAPL}$ only looks at the l_1 distances between the two distributions the point-wise distance can be rather significant. Thus, the privacy of any individual may be compromised, without the knowledge of whether the data vector happens to belong to the set with unusually high distance between the two distributions. Though the probability of this event may be low, it is precisely the privacy of outliers that we ought to protect. Another point worth noting is the following – our assumption that the data records is generated from IID random variables is may not be entirely realistic in some practical applications. We hope to relax this assumption in our future research.

Acknowledgements

Vincent Tan is supported by A*STAR, Singapore. This work was performed at the Institute for Infocomm Research (I^2R). This paper also has benefited from discussions with Dr. Mafruzzaman Ashrafi from I^2R.

References

1. Agrawal, D., Aggarwal, C.C.: On the design and quantification of privacy preserving data mining algorithms. In: Proc. of Symposium on Principles of Database Systems, pp. 247–255 (2001)
2. Agrawal, R., Srikant, R.: Privacy-preserving data mining. In: In Proc. of ACM SIGMOD Conf. on Management of Data, pp. 439–450 (2000)
3. Atallah, M., Bertino, E., Elmagarmid, A., Ibrahim, M., Verykios, V.: Disclosure limitation of sensitive rules. In: Proc. of Knowledge and Data Engineering Exchange, 1999 (KDEX 1999), pp. 45–52 (1999)
4. Ben-Or, M., Goldwasser, S., Wigderson, A.: Completeness theorems for cryptographic fault-tolerant distributed computation. In: Proc. of 20th ACM Symposium on the Theory of Computation (STOC), pp. 1–10 (1988)
5. Caetano, T.: Graphical Models and Point Set Matching. PhD thesis, Universidade Federal do Rio Grande do Sul (UFRGS) (2004)
6. Chang, C.C., Lin, C.J.: LIBSVM: A library for support vector machines (2001), http://www.csie.ntu.edu.tw/~cjlin/libsvm
7. Chen, K., Liu, L.: Privacy preserving data classification with rotation perturbation. In: Proc. of 5th IEEE Int. Conf. on Data Mining (ICDM 2005), Houston, TX, pp. 589–592 (2005)
8. Devroye, L.: Sample-based non-uniform random variate generation. In: 18th conference on Winter simulation (1985)
9. Devroye, L.: Non-Uniform Random Variate Generation. Springer, New York (1986)
10. Devroye, L., Gyorfi, L.: Non-parametric Density Estimation. The L1 View. Wiley, Chichester (1955)
11. Du, W., Han, Y.S., Chen, S.: Privacy-preserving multivariate statistical analysis: Linear regression and classification. In: Jonker, W., Petković, M. (eds.) SDM 2004. LNCS, vol. 3178, pp. 83–99. Springer, Heidelberg (2004)
12. Duda, R.O., Hart, P.E., Stork, D.G.: Pattern Classification. Wiley, Chichester (2000)
13. Dwork, C., Kenthapadi, K., McSherry, F., Mironov, I., Naor, M.: Our Data, Ourselves: Privacy Via Distributed Noise Generation. In: Vaudenay, S. (ed.) EUROCRYPT 2006. LNCS, vol. 4004, pp. 486–503. Springer, Heidelberg (2006)
14. Evfimievski, A.: Randomization in privacy preserving data mining. ACM SIGKDD Explorations Newsletter 4, 43–48 (2002)
15. Evfimievski, A., Srikant, R., Agrawal, R., Gehrke, J.: Privacy preserving mining of association rules. In: Proc. of 8th ACM SIGKDD Int. Conf. on Knowledge Discovery in Databases and Data Mining, pp. 217–228 (2002)
16. Fukunaga, K., Hostetler, L.D.: The estimation of gradient of a density function with applications to pattern recognition. IEEE Transactions on Information Theory 21, 32–40 (1975)
17. Han, J., Kamber, M.: Data Mining: Concepts and Techniques. Morgan Kaufmann Publishers, San Francisco (2000)
18. Huang, Z., Du, W., Chen, B.: Deriving private information from randomized data. In: Proc. of ACM SIGMOD Conf., Baltimore, MD, pp. 37–48 (2005)
19. Indyk, P., Woodruff, D.: Polylogarithmic private approximations and efficient matching. In: Proc. of Theory of Cryptography Conf., NY (2006)
20. Kargupta, H., Datta, S., Wang, Q., Sivakumar, K.: On the privacy preserving properties of random data perturbation techniques. In: Proc. of 3rd IEEE Int. Conf. on Data Mining, Washington, DC, USA, pp. 99–106 (2003)

21. Kargupta, H., Park, B., Hershbereger, D., Johnson, E.: Collective data mining: A new perspective toward distributed data mining. In: Advances in distributed data mining, pp. 133–184 (1999)

22. Liew, C.K., Choi, U.J., Liew, C.J.: A data distortion by probability distribution. ACM Trans. Database Systems (TODS) 10, 395–411 (1985)

23. Lindell, Y., Pinkas, B.: Privacy Preserving Data Mining. In: Bellare, M. (ed.) CRYPTO 2000. LNCS, vol. 1880, pp. 36–53. Springer, Heidelberg (2000)

24. Liu, K., Kargupta, H., Ryan, J.: Random projection-based multiplicative data perturbation for privacy preserving distributed data mining. IEEE Transactions on Knowledge and Data Engineering (TKDE) 18, 92–106 (2006)

25. Merugu, S., Ghosh, J.: A privacy-sensitive approach to distributed clustering. Special issue: Advances in pattern recognition 26(4), 399–410 (2005)

26. Muralidhar, K., Parsa, R., Sarathy, R.: A general additive data perturbation method for database security. Management Science 19, 1399–1415 (1999)

27. Newman, D.J., Hettich, S., Blake, C.L., Merz, C.J.: UCI Repository of Machine Learning Databases, University of California, Irvine, Dept. of Information and Computer Sciences (1998),
http://www.ics.uci.edu/~mlearn/MLRepository.html

28. Oliveira, S.R., Zaiane, O.R.: A privacy-preserving clustering approach toward secure and effective data analysis for business collaboration. Computers & Security 26(1), 81–93 (2007)

29. Parzen, E.: On the estimation of a probability density function and mode. Annals of Mathematical Statistics 33, 1065–1076 (1962)

30. Pinkas, B.: Cryptographic techniques for privacy preserving data mining. SIGKDD Explorations 4, 12–19 (2002)

31. Samarati, P., Sweeney, L.: Protecting privacy when disclosing information: k-anonymity and its enforcement through generalization and suppression. In: Proc. of the IEEE Symposium on Research in Security and Privacy, Oakland, CA (May 1998)

32. Scott, D.W.: Multivariate Density Estimation. Theory, Practice and Visualization. Wiley, Chichester (1992)

33. Silverman, B.W.: Density Estimation for Statistics and Data Analysis. Chapman & Hall, London (1986)

34. Subramaniam, H., Wright, R.N., Yang, Z.: Experimental analysis of privacy-preserving statistics computation. In: Proc. of the Workshop on Secure Data Management (in conjunction with VLDB 2004) (2004)

35. Sweeney, L.: k-anonymity: A model for protecting privacy. Int. Journal of Uncertainty Fuzziness Knowledge Based Systems 10, 557–570 (2002)

36. Vaidya, J., Clifton, C.: Privacy preserving Naïve Bayes classifier for vertically partitioned data. In: Jonker, W., Petković, M. (eds.) SDM 2004. LNCS, vol. 3178, pp. 330–334. Springer, Heidelberg (2004)

37. Yao, A.: How to generate and exchange secrets. In: Proc. 27th IEEE Symposium on Foundations of Computer Science, pp. 162–167 (1986)

38. Zhang, N., Wang, S., Zhao, W.: A new scheme on privacy-preserving data classification. In: Proc. of 11th ACM SIGKDD Int. Conf. on Knowledge Discovery in Data Mining, pp. 374–383 (2005)

A   Proofs of Lemmas 4.1 and 4.2

Proof. Since random projection-based multiplicative data perturbation method is orthogonal on expectation [24], $\mathbb{E}\left[\mathbf{Y}_{(1)}^T \mathbf{Y}_{(2)}\right] = \mathbf{x}_{(1)}^T \mathbf{x}_{(2)}$, the columns are also orthogonal on expectation i.e. $\mathbb{E}\left[\tilde{\mathbf{Y}}_{(1,i_1)}^T \tilde{\mathbf{Y}}_{(2,i_2)}\right] = \tilde{\mathbf{x}}_{(1,i_1)}^T \tilde{\mathbf{x}}_{(2,i_2)}$ for all $1 \leq i_1, i_2 \leq d$. Using the Cauchy-Schwarz Inequality, we have

$$\tilde{\mathbf{x}}_{(1,i_1)}^T \tilde{\mathbf{x}}_{(2,i_2)} \leq \|\tilde{\mathbf{x}}_{(1,i_1)}\| \|\tilde{\mathbf{x}}_{(2,i_2)}\|. \tag{21}$$

Since we are also given $\|\widetilde{\tilde{\mathbf{x}}_{(2,i_2)}}\|$, we can bound $\|\tilde{\mathbf{x}}_{(2,i_2)}\|$,

$$\|\tilde{\mathbf{x}}_{(1,i_1)}\| \geq \frac{\mathbb{E}\left[\tilde{\mathbf{Y}}_{(1,i_1)}^T \tilde{\mathbf{Y}}_{(2,i_2)}\right]}{\|\widetilde{\tilde{\mathbf{x}}_{(2,i_2)}}\|} \triangleq \gamma_1, \tag{22}$$

This yields Eq (7). Lemma 4.2 follows directly from Lemma 1 by subtracting elements contained in the set $\mathcal{A}_{i_1, \backslash q}$ as defined in Eqn (8).   □

Towards Privacy-Preserving Model Selection*

Zhiqiang Yang[1], Sheng Zhong[2], and Rebecca N. Wright[3]

[1] Department of Computer Science, Stevens Institute of Technology,
Hoboken, NJ, 07030, USA
zyang@cs.stevens.edu
[2] Department of Computer Science, SUNY Buffalo, Buffalo, NY, 14260, USA
szhong@cse.buffalo.edu
[3] Department of Computer Science and DIMACS, Rutgers University,
Piscataway, NJ, 08854, USA
rebecca.wright@rutgers.edu

Abstract. Model selection is an important problem in statistics, machine learning, and data mining. In this paper, we investigate the problem of enabling multiple parties to perform model selection on their distributed data in a privacy-preserving fashion without revealing their data to each other. We specifically study cross validation, a standard method of model selection, in the setting in which two parties hold a vertically partitioned database. For a specific kind of vertical partitioning, we show how the participants can carry out privacy-preserving cross validation in order to select among a number of candidate models without revealing their data to each other.

1 Introduction

In today's world, a staggering amount of data, much of it sensitive, is distributed among a variety of data owners, collectors, and aggregators. Data mining provides the power to extract useful knowledge from this data. However, privacy concerns may prevent different parties from sharing their data with others. A major challenge is how to realize the utility of this distributed data while also protecting data privacy.

Privacy-preserving data mining provides data mining algorithms in which the goal is to compute or approximate the output of one or more particular algorithms applied to the joint data, without revealing anything else, or at least anything else sensitive, about the data.

To date, work on distributed privacy-preserving data mining has been primarily limited to providing privacy-preserving versions of particular data mining algorithms. However, the data miner's task rarely starts and ends with running a particular data mining algorithm. In particular, a data miner seeking to model some data will often run a number of different kinds of data mining algorithms

---

* This work was supported in part by the National Science Foundation under Grant No. CCR-0331584 and by the Department of Homeland Security under ONR Grant N00014-07-1-0159.

F. Bonchi et al. (Eds.): PinKDD 2007, LNCS 4890, pp. 138–152, 2008.

and then perform some kind of model selection to determine which of the resulting models to use. If privacy-preserving methods are used for determining many models, but then the model selection either is carried out without maintaining privacy or cannot be carried out due to privacy constraints, then the desired privacy and utility cannot simultaneously be achieved.

In this paper, we introduce the notion of privacy-preserving model selection. We specifically consider cross validation, which is a popular method for model selection. In cross validation, a number of different models are generated on a portion of the data. It is then determined how well the resulting models perform on the rest of the data, and the highest performing model is selected. Cross validation can also be used to validate a single model learned from training data on test data not used in the generation of the model, to determine whether the model performs sufficiently well and limit the possibility of choosing a model that overfits the data.

We provide a partial solution to privacy-preserving model selection via cross validation. We assume a very specific kind of vertical partitioning, which has previously been considered by Vaidya and Clifton [31], in which one party holds all the data except the class labels, and a second party holds all the class labels. In this setting, we show how to perform model selection using cross validation in a privacy-preserving manner, without revealing the parties' data to each other.

A practical example of this kind of partitioning might occur in a research project seeking to explore the relationship between certain criminal activities and the medical histories of people involved in these activities. A local hospital has a database of medical histories, while the police department has the criminal records. Both the hospital and police department would like to provide assistance to this project, but neither of them is willing or legally able to reveal their data to the other. It is therefore a technical challenge to find the right model over this distributed database in a privacy-preserving manner. Specifically, we can view the medical histories and the criminal records as two parts of a vertically partitioned database. We simplify the criminal records to labels on local residents for whether they are involved in the criminal activities. Then the question becomes finding the right model to predict this label on an individual using his or her medical history data. We require that the medical histories not be revealed to the police department and that the labels not be revealed to the hospital.

Our main contribution is a privacy-preserving model selection protocol for this setting. Specifically, there is a database vertically partitioned between two participants. One participant has all the data except the class labels; the other participant has all the class labels. Our privacy-preserving cross validation solution enables the parties to privately determine the best among a number of candidate models for the data, thereby extending the privacy of the data from the initial model computation through to the model selection step.

We begin by discussing related work in Section 2. We introduce some cryptographic primitives in Sections 3 and 4. Our main protocol is shown in Section 5. In Section 6, we discuss possible extensions including generalizing our solution to arbitrary vertically partitioned data and determining which of a number of models is best without revealing the models that are not chosen.

2   Related Work

Existing techniques in privacy-preserving data mining can largely be categorized into two approaches. One approach adopts cryptographic techniques to provide secure solutions in distributed settings, as pioneered by Lindell and Pinkas [25]. Another approach randomizes the original data such that certain underlying patterns are still kept in the randomized data, as pioneered by Agrawal and Srikant [3]. Generally, the cryptographic approach can provide solutions with perfect accuracy and perfect privacy. The randomization approach is much more efficient than the cryptographic approach, but typically suffers a tradeoff between privacy and accuracy.

In the randomization approach, original data is randomized by adding noise so that the original data is disguised but patterns of interest persist. The randomization approach enables learning data mining models from the disguised data. Different randomization approaches have been proposed to learn different data mining models, such as decision trees [3,8] and association rules [28,11,10]. Several privacy definitions for the randomization setting have been proposed to achieve different levels of privacy protection [3,1,10], though privacy problems with randomization approach have also been discussed [23,17].

In the cryptographic approach, which we follow in this paper, the goal is to enable distributed learning of data mining models across different databases without the database owners revealing anything about their data to each other beyond what can be inferred from the final result. In principle, general-purpose secure multiparty computation protocols [16,35] can provide solutions to any distributed privacy-preserving data mining problem.

However, these general-purpose protocols are typically not efficient enough for use in practice when the input data is very large. Therefore, more efficient special-purpose privacy-preserving protocols have been proposed for many special cases. These address a number of different learning problems across distributed databases, such as association rule mining [29,21], ID3 trees [25], clustering [30,19], naive Bayes classification [22,31], and Bayesian networks [27,34], as well as a variety of privacy-preserving primitives for simple statistical computations including scalar product [7,4,29,33,13,14], finding common elements [13,2], and computing correlation matrices [26].

In the cryptographic approach, privacy is quantified using variants of the standard cryptographic definition for secure multiparty computation. Intuitively, parties involved in the privacy-preserving distributed protocols should learn only the data mining results that are their intended outputs, and nothing else.

Most privacy-preserving data mining solutions to date address typical data mining algorithms, but do not address necessary preprocessing and postprocessing steps. Recent work addresses privacy preservation during the preprocessing step [20]. In this work, we begin the exploration of extending the preservation of privacy to the postprocessing step, thereby maintaining privacy throughout the data mining process.

3   Cryptographic Tools

In this section, we briefly overview cryptographic concepts and primitives that we use.

3.1   Privacy Definition

In this paper, we define privacy using a standard definition used in secure multiparty computation [15]. In particular, we consider the privacy definition in the model of semi-honest adversaries. A semi-honest adversary is assumed to follow its specified instructions, but will try to gain as much information as possible about other parties' inputs from the messages it receives.

The proofs of privacy in this paper are carried out using the simulation paradigm [15]. Formally, let Π be a 2-party protocol for computing a function $f : (x_1, x_2) \rightarrow (y_1, y_2)$. The *view* of the ith party ($i \in \{1, 2\}$) during an execution of Π, denoted by $\text{view}_i(x_1, x_2)$, consists of the ith party's input x_i, all messages received by the ith party, and all internal coin flips of the ith party. We say that Π *privately computes f against semi-honest adversaries* if, for each i, there exists a probabilistic polynomial-time algorithm S_i (which is called a *simulator*), such that

$$\{S_i(x_i, y_i)\}_{x_1, x_2} \stackrel{c}{\equiv} \{(\text{view}_i(x_1, x_2)\}_{x_1, x_2},$$

where $\stackrel{c}{\equiv}$ denotes *computational indistinguishability*. (See [15] for the definition of computational indistinguishability. Intuitively, it states that a polynomially-bounded computation cannot distinguish between the two distributions given samples from them.)

Intuitively, this definition states that, based on the input and output of each participant, we should be able to "simulate" the view of that participant. Therefore, each participant learns nothing during the computation that would not be learned if Alice and Bob gave their data to a trusted third party who computed the results y_1 and y_2 and returned them to Alice and Bob, respectively.

As a privacy definition, this definition has some advantages but it also has some limitations. Among its advantages are that it allows provable guarantees that nothing was leaked during the computation, and that if multiple subprotocols are combined properly, their combination does not leak any information. A notable limitation of this definition is that it does not say anything about the privacy of the final result, leaving that determination as a separate privacy decision.

3.2   ElGamal Cryptosystem

A public key cryptosystem consists of three algorithms: the *key generation algorithm*, the *encryption algorithm*, and the *decryption algorithm*. We make use of the ElGamal cryptosystem [9].

In the ElGamal cryptosystem, the key generation algorithm generates the *parameters* (G, q, g, x), where G is a cyclic group of order q with generator g, and x is randomly chosen from $\{0, \ldots, q-1\}$. The *public key* is (h, G, q, g) where $h = g^x$, and the *private key* is x.

In order to encrypt a message m to Alice under her public key (h, G, q, g), Bob computes $(c_1 = m \cdot h^k, c_2 = g^k)$, where k is randomly chosen from $\{0, \ldots, q-1\}$. To decrypt a ciphertext (c_1, c_2) with the private key x, Alice computes $c_1(c_2^x)^{-1}$ as the plaintext message.

ElGamal encryption is *semantically secure* under the Decisional Diffie-Hellman (DDH) assumption [5], which we assume throughout this paper. One group family in which DDH is commonly assumed to be intractable is the quadratic residue subgroup of $\mathbb{Z}_p^*$ (the multiplicative group mod p) where p is a *safe* prime (i.e., a prime number of the form $p = 2p' + 1$ for a prime p'). ElGamal encryption has a useful *randomization property*. Specifically, given an encryption of M, it is possible to compute a different (and random) encryption of M without knowing the private key.

4   Privacy-Preserving Hamming Distance and Generalized Hamming Distance

In this section, we provide new, simple, efficient, privacy-preserving protocols for computing the Hamming distance and generalized Hamming distance between two vectors. These will be used in our main protocol.

4.1   Privacy-Preserving Hamming Distance

In this protocol, Alice has a vector $A = (a_1, ..., a_n)$ and Bob has a vector $B = (b_1, ..., b_n)$, where A and B contain only binary values. In our setting, Alice is supposed to learn the Hamming distance of their two vectors, and Alice and Bob are supposed to learn nothing else about each other's vectors. (In the semi-honest setting, such a protocol can easily be transformed into a protocol where both Alice and Bob learn the result, by having Alice tell Bob the answer.)

We note that private solutions already exist for this problem. For example, an efficient solution is given by Jagannathan and Wright [20] based on homomorphic encryption. Yao's secure two-party computation could be used [35]; it computes the result based on computation using a "garbled" circuit. Alternately, the secure two-party computation techniques of Boneh, Goh, and Nissim [6] could be used, in the form where the output is multiple bits; this relies on computationally expensive bilinear pairing, as well as on a new computational assumption. We also note that if one is willing to accept an approximation to the Hamming distance, it is possible to achieve this with sublinear communication complexity while meeting the privacy requirements [12, 18].

In this section, we describe a simple, efficient, alternative solution based on the ElGamal cryptosystem. As shown in the following section, a modification of this solution also solves the generalized Hamming distance problem.

Privacy-Preserving Hamming Distance Protocol

Input: Vectors $A = (a_1, \ldots, a_n)$ and $B = (b_1, \ldots, b_n)$ held by Alice and Bob, respectively, such that $a_i, b_i \in \{0, 1\}$ for $1 \leq i \leq n$.

Output: Alice learns the Hamming distance of A and B.

1. For $1 \leq i \leq n$, if $a_i = 0$, then Alice sends $e_i = E(g)$ to Bob; Otherwise, Alice sends $e_i = E(g^{-1})$ to Bob. (Obviously, fresh, independent randomness is used in generating each of these encryptions.) For each i, the resulting encryption is a two-part ciphertext $e_i = (c_{i,1}, c_{i,2})$.
2. For $1 \leq i \leq n$, if $b_i = 0$, then Bob rerandomizes e_i to get e_i'; otherwise, Bob sets e_i' to a rerandomization of $e_i^{-1} = ((c_{i,1})^{-1}, (c_{i,2})^{-1})$. Bob sends a permuted vector including all e_i' to Alice.
3. Alice decrypts all received e_i' and counts how many of the decryptions are equal to g^{-1}. This number is equal to $\sum_{i=1}^{n}(a_i \oplus b_i) = \text{dist}(A, B)$.

Fig. 1. Privacy-Preserving Hamming Distance Protocol

In our protocol, we assume Alice has an ElGamal key pair (x, y) ($x \in [0, q-1]$, where q is the order of G; $y \in G$) such that $y = g^x \in G$. Here, x is the private key, which is known only to Alice, and y is the public key, which is also known to Bob. We use $E(m)$ to denote an encryption of m by public key y. All computations in the protocol and throughout this paper take place in G, which is chosen large enough to ensure that the final distance result is correct as an integer. The output of this protocol is the Hamming distance $\text{dist}(A, B) = \sum_{i=1}^{n}(a_i \oplus b_i)$. The basic idea is that we use g to represent 0 and g^{-1} to represent 1. The protocol is shown in Figure 1.

Theorem 1. *Under the DDH assumption, the protocol in Figure 1 for binary-valued inputs privately computes the Hamming distance in the semi-honest model.*

Proof. We first show correctness—i.e., that Alice's output is the correct Hamming distance. In Step 1, for $1 \leq i \leq n$, Alice computes

$$e_i = (c_{i,1}, c_{i,2}) = \begin{cases} E(g) & \text{if } a_i = 0 \\ E(g^{-1}) & \text{if } a_i = 1 \end{cases}$$

and sends these to Bob. In Step 2, for $1 \leq i \leq n$, Bob produces e_i'. If $b_i = 0$, then e_i' is a rerandomization of the encryption e_i. In this case, e_i' encrypts the same cleartext g or g^{-1} that e_i does (even though Bob does not himself know this cleartext). If, on the other hand, $b_i = 1$, then Bob sets e_i' to be a rerandomization of $(c_{i,1}{}^{-1}, c_{i,2}{}^{-1})$. Assuming k_i is the random value used in Alice's encryption of $m_i \in \{g, g^{-1}\}$, then:

$$(c_{i,1}{}^{-1}, c_{i,2}{}^{-1}) = ((m_i \cdot h^{k_i})^{-1}, (g^{k_i})^{-1}) = (m_i{}^{-1} \cdot h^{-k_i}, g^{-k_i}).$$

Once rerandomized so as to use fresh randomization, this is a valid encryption of $m_i{}^{-1}$. Thus, if $m_i = g$, then e_i' is an encryption of g^{-1}, and if $m_i = g^{-1}$, then e_i' is an encryption of g. It further follows that e_i' is an encryption of g^{-1} if and only if $a_i \neq b_i$. (If $a_i = b_i$, then e_i' is an encryption of g.) Therefore, the number of g^{-1} decryptions Alice obtains in Step 3 is the desired Hamming distance.

To show privacy, we need to demonstrate simulators S_1 for Alice and S_2 for Bob. We construct S_1 as follows. S_1 simulates all internal coin flips of Alice as described in the protocol. S_1 simulates the message from Bob to Alice using a randomly permuted vector of n ElGamal ciphertexts; among these n ciphertexts, the number of encryptions of g^{-1} should be equal to the output of Alice; all the remaining ciphertexts should be encryptions of g.

We construct S_2 as follows. S_2 simulates all internal coin flips of Bob as described in the protocol. S_2 simulates the message from Alice to Bob using n random ElGamal ciphertexts. The computational indistinguishability immediately follows from the semantic security of the ElGamal cryptosystem under the DDH assumption.

4.2   Privacy-Preserving Generalized Hamming Distance

In this protocol, we consider the case that $A = (a_1, \ldots, a_n)$ and $B = (b_1, \ldots, b_n)$ where each a_i and each b_i has a finite domain $\{1, \ldots, s\}$ rather than a binary domain. For these general discrete-valued a_i and b_i, we consider the Boolean difference function:

$$\mathsf{diff}(a_i, b_i) = \begin{cases} 0 \text{ if } a_i = b_i \\ 1 \text{ otherwise.} \end{cases}$$

We define the *generalized Hamming distance* as $\mathsf{gdist}(A, B) = \sum_{i=1}^{n} \mathsf{diff}(a_i, b_i)$. Our protocol for privately computing generalized Hamming distance, shown in Figure 2. Like the Hamming distance protocol, the generalized Hamming distance protocol also relies on the ElGamal cryptosystem. In this case, we take advantage of homomorphic properties obtained by encrypting in the exponent. That is, we encrypt a message m by using g^m as the cleartext in the ElGamal system rather than using m.

Theorem 2. *Under the DDH assumption, the protocol in Figure 2 for general discrete-valued inputs privately computes the generalized Hamming distance in the semi-honest model.*

Privacy-Preserving Generalized Hamming Distance Protocol

Input: Vectors $A = (a_1, \ldots, a_n)$ and $B = (b_1, \ldots, b_n)$ held by Alice and Bob, respectively, such that $a_i, b_i \in \{1, \ldots, s\}$ for $1 \leq i \leq n$.

Output: Alice learns gdist(A, B).

1. For $1 \leq i \leq n$, Alice sends $E(g^{a_i}) = (c_{i,1}, c_{i,2})$ to Bob.
2. For $1 \leq i \leq n$, Bob computes $E(g^{b_i}) = (c'_{i,1}, c'_{i,2})$ and defines $e'_i = (c'_{i,1}/c_{i,1}, c'_{i,2}/c_{i,2}) = (d_{i,1}, d_{i,2})$. Then Bob chooses a random r_i and computes $e''_i = ((d_{i,1})^{r_i}, (d_{i,2})^{r_i})$. Bob sends a random permutation of all the e''_i to Alice.
3. Alice decrypts all received e''_i. Alice counts the total number of decryptions whose values are not equal to 1. This number is equal to $\sum_{i=1}^{n} \text{diff}(a_i, b_i) = \text{gdist}(A, B)$.

Fig. 2. Privacy-Preserving Generalized Hamming Distance Protocol

Proof. We begin by showing correctness. For $1 \leq i \leq n$, for some random values k_i and ℓ_i, we have:

$$
\begin{aligned}
e''_i &= (d_{i,1}{}^{r_i}, d_{i,2}{}^{r_i}) \\
&= \left(\frac{c'_{i,1}{}^{r_i}}{c_{i,1}{}^{r_i}}, \frac{c'_{i,2}{}^{r_i}}{c_{i,2}{}^{r_i}} \right) \\
&= \left(\frac{g^{b_i} \cdot h^{k_i}}{g^{a_i} \cdot h^{\ell_i}}, \frac{g^{k_i}}{g^{\ell_i}} \right) \\
&= (g^{(b_i - a_i)} \cdot h^{(k_i - \ell_i)}, g^{(k_i - \ell_i)}).
\end{aligned}
$$

Thus, e''_i decrypts to $g^{(b_i - a_i)}$, which is equal to 1 if and only if $a_i = b_i$. It follows that the number Alice obtains in Step 3 of decryptions that are not equal to 1 is the desired distance.

To show privacy, we must show simulators S_1 for Alice and S_2 for Bob. S_1 simulates all internal coin flips of Alice as described in the protocol. S_1 simulates the message from Bob to Alice using a randomly permuted vector of n ElGamal ciphertexts. These n ciphertexts are chosen so that among these n ciphertexts, the number of encryptions of 1 is equal to the output of Alice; the remaining ciphertexts are encryptions of random cleartexts.

S_2 simulates all internal coin flips of Bob as described in the protocol. S_2 simulates the message from Alice to Bob using n random ElGamal ciphertexts.

The computational indistinguishability immediately follows from the semantic security of the ElGamal cryptosystem under the DDH assumption.

4.3 Experimental Results

We implemented these privacy-preserving Hamming distance and generalized Hamming distance protocols using the OpenSSL library in C. We carried out our

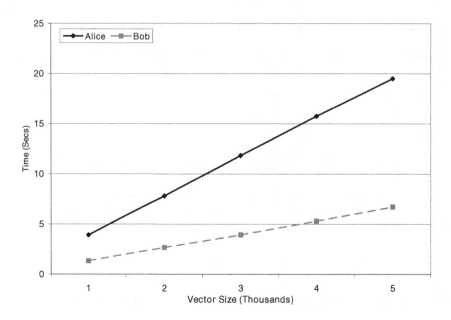

Fig. 3. Performance of Privacy-Preserving Hamming Distance Protocol

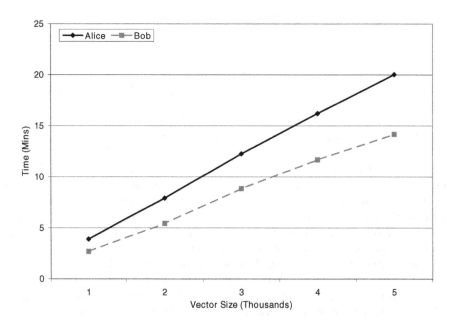

Fig. 4. Performance of Privacy-Preserving Generalized Hamming Distance Protocol

experiments on a NetBSD machine with 2GHz CPU and 512M memory, using public keys of 1024 bits. The computation cost dominates the overall protocol, so we measured only the computation time for Alice and Bob. (That is, we did not measure the communication time.) We measured the computation cost of both the Hamming distance protocol and the generalized Hamming distance protocol on binary-valued vectors of varying lengths.

The results of our experiments are shown in Figure 3 for the Hamming distance protocol and Figure 4 for the generalized Hamming distance protocol. As expected, the experiments demonstrate that the computation time of Alice and Bob is linear in the vector size for both protocols. For the same length vectors, the generalized Hamming distance protocol takes Bob about twice as long to compute as the Hamming distance protocol.

5   Privacy-Preserving Model Selection

Many models have been proposed in the field of statistics, machine learning, and data mining, including linear models, neural networks, classification and regression trees, and kernel methods. One of the problems in data mining is how to select which kind of model is best for a particular task in a particular setting. Often, a human expert will make some initial judgment about which model or models seem likely candidates for the task at hand. With or without expert guidance, it is very common and useful to use measure the performance of a few learned models on test data to see which performs the best. Model selection is also useful for determining the parameters to use for a particular model, such as the depth of a decision tree.

In the setting of privacy-preserving data mining, it is important that the privacy that was maintained in learning data mining models is not lost in the model selection process. As mentioned earlier, in this paper we provide a first step towards a solution. We address a specific kind of vertical partitioning, in which one party holds all the data except the class labels and a second party holds all the class labels.

5.1   Problem Formulation

A database D is vertically partitioned between two parties, Alice and Bob. D contains n records in total. Alice's database D_A consists of m non-class attributes $(V_1, ..., V_m)$, where each V_i has a finite domain. Bob's database D_B includes only the class attribute C.

The parties want to collaborate to select an appropriate classification model to learn based on the combination of their databases—e.g., to decide whether to learn decision trees or naive Bayes classifiers from their data. However, because of privacy concerns, they do not wish to reveal their original data to each other.

Cross validation is a popular method for model selection. To carry out cross validation in a privacy-preserving manner, it is necessary to prevent the parties from learning anything they would not otherwise learn. In our setting, this means that Bob should not learn the predicted class labels for various models applied to Alice's data, and Alice should not learn the class labels that Bob has. In addition, candidate models should themselves be generated in a privacy-preserving way.

We present a privacy-preserving protocol for selecting a model on a database vertically partitioned so that Bob holds only the class labels. Specifically, we present a privacy-preserving protocol for selecting a model between decision trees and naive Bayes classifiers. Our protocol easily extends to any kind of classifier that can be learned in a privacy-preserving manner.

Privacy-preserving protocols for learning decision trees and for learning naive Bayes classifiers, respectively, have been proposed by Vaidya and Clifton [32,31]. Here, we use these two protocols as "black boxes" to achieve privacy-preserving model selection, using k-fold cross validation as an example. Our results extend straightforwardly to any type of model selection in which the choice of model depends only on the available data and the number of errors made by each model. This includes other types of cross validation such as the holdout method and leave-one-out cross validation.

In k-fold cross validation, both parties partition (their own parts of) the original database D into k pieces. The first $k-1$ pieces are *training sets* used to learn the model and the remaining piece is the *test set* used to validate or test the learned models. The parties learn both types of candidate model using a privacy-preserving protocol on each of the training sets (resulting in $k-1$ decision trees and $k-1$ naive Bayes classifiers). They then use the training set for estimating the classification error of each type of learned model. The type of model (i.e., decision tree or naive Bayes classifier) that has the lowest mean classification error over all $k-1$ learned models is considered the best, and is then learned (presumably again in a privacy-preserving manner) over the entire dataset. To compute the mean classification error for each type of model, we show how to compute the classification error for a single model in a privacy-preserving manner.

5.2   Our Protocol

After multiple classification models are learned from the database D for each of $k-1$ training sets, both parties need to compute the classification error for each model on the test set. Because Alice has all the non-class attributes, she can classify each record by herself using the classification model which has been learned. However, the classification error depends on the real class label which is held by Bob.

For privacy reasons, Alice cannot send the classification results to Bob, and Bob cannot send the actual class labels to Alice, as this would breach their privacy. Similarly, Alice cannot send her data to Bob so that he can apply the classifiers on her data and compare the results to the actual class labels. To get around this, we instead compute the classification errors using a privacy-preserving Hamming distance protocol (or its generalized version, if there are more than two possible class labels) such as the one presented in Section 3.

Alice's input to each instance of Hamming distance protocol is a vector $A = (a_1, ..., a_n)$, where each a_i is the class label predicted by Alice using the learned

---

Privacy-Preserving k-fold Cross Validation

Input: A database D vertically partitioned between Alice and Bob. Bob holds the class attribute and Alice holds all the other attributes.

Output: Alice and Bob learn the selected model for D.

1. Alice and Bob partition the database D into k pieces ($k - 1$ training sets and one test set).
2. Alice and Bob use privacy-preserving protocols on the $k - 1$ training sets to learn $k - 1$ decision trees and $k - 1$ naive Bayes classifiers.
3. For $1 \leq i \leq k - 1$, Alice and Bob carry out the following steps:
   (a) Alice classifies her records in the test set using the ith learned decision tree and naive Bayes classifier.
   (b) Alice and Bob use the privacy-preserving Hamming distance protocol (or generalized Hamming distance protocol, as appropriate) for Alice to compute the classification error from the ith learned decision tree and from the ith learned naive Bayes classifier on the test set. (That is, comparing her results computed in Step 3a to Bob's actual class values in the test set.)
4. Combining the $k - 1$ results, Alice computes the mean classification error for the decision tree and for the naive Bayes classifier, and announces these results to Bob. Alice and Bob then select the type of model which has the lower mean classification error.

Fig. 5. Privacy-Preserving Model Selection Protocol

classification model, where n is the number of records in the test set. Bob's input is a vector $B = (b_1, ..., b_n)$, where each b_i is the real class label. From the resulting classification errors, Alice can determine the mean classification error for each type of model. We summarize the entire protocol in Figure 5.

As we have described it in Figure 5, the protocol leaks partial information in step 3b—namely, the number of misclassified records for each model on the test set. However, if desired, one could remove this leak by stopping all of the generalized Hamming distance protocols before Bob sends the the final results to Alice, and using a Yao secure two-party computation on Bob's encryptions and Alice's decryption key for the parties to learn only the mean classification error. However, this step would add substantial additional cost unless the total number of records in the test set and total number of training sets are relatively small.

6   Discussion

In this paper, we introduced privacy-preserving model selection. This is important for "extending the boundary" of privacy-preserving protocols to include steps beyond the computation of particular data mining models. By extending the boundary of what can be accomplished with efficient privacy-preserving computation, we bring the adoption of privacy-preserving data mining closer to practice.

Our privacy-preserving solution enables model selection via cross validation on a database vertically partitioned between two parties. Our solution is based on a privacy-preserving primitive for computing the Hamming distance or generalized Hamming distance of two vectors, which may be of independent interest.

There are a number of interesting directions yet to be studied. Most importantly, our setting considers a very extreme case of partitioning. We argue this is realistic in some cases, but clearly it is not applicable in all cases. Additionally, in order to provide the greatest privacy protection during the model selection process not only the raw data, but also the candidate models that are not selected should be kept private, as revealing multiple models provides more information about the data than revealing just the selected type of model does.

An attractive option, that could both allow more general vertical partitioning and protect privacy further by not releasing the individual candidate models, is to have the models themselves be computed in such a way that they are not known to the individual parties, but can be used by them through yet another protocol. To our knowledge, only a small number of privacy-preserving protocols do this—namely, Vaidya and Clifton's naive Bayes classifier protocol [31] and Laur, Lipmaa and Mielikäinen's support vector machines [24]. By using such protocols and modifying our Hamming distance protocol to have Alice's input shared by Alice and Bob instead of known to Alice, it should be possible to obtain a solution in which the parties perform model selection without revealing the candidate models considered. This kind of solution would also be appealing because it can maintain the privacy of the classifier results even in the case that both parties know the real class labels.

Our proposed protocol defends against semi-honest adversaries. It is open to extend them to efficient protocols that provide security against malicious adversaries. It might also be interesting to consider other distance metrics such as the L1-distance that allow for finer granularity by considering some wrong answers more acceptable than others. Finally, as the privacy definitions in secure multiparty computation are very strict, relaxed yet meaningful privacy definitions that enable more practical protocols deserve further exploration.

Acknowledgments

We thank the attendees of the PinKDD'07 workshop for helpful and interesting discussions.

References

1. Agrawal, D., Aggarwal, C.: On the design and quantification of privacy preserving data mining algorithms. In: Proc. of the 20th ACM SIGMOD-SIGACT-SIGART Symposium on Principles of Database Systems, pp. 247–255 (2001)
2. Agrawal, R., Evfimievski, A., Srikant, R.: Information sharing across private databases. In: Proc. of the 2003 ACM SIGMOD International Conference on Management of Data, pp. 86–97 (2003)

3. Agrawal, R., Srikant, R.: Privacy preserving data mining. In: Proc. of the 2000 ACM SIGMOD International Conference on Management of Data, pp. 439–450 (May 2000)
4. Atallah, M., Du., W.: Secure multi-party computational geometry. In: Proc. of the Seventh International Workshop on Algorithms and Data Structures, pp. 165–179. Springer, Heidelberg (2001)
5. Boneh, D.: The decision Diffie-Hellman problem. In: Buhler, J.P. (ed.) ANTS 1998. LNCS, vol. 1423, pp. 48–63. Springer, Heidelberg (1998)
6. Boneh, D., Goh, E., Nissim, K.: Evaluating 2-DNF formulas on ciphertexts. In: Kilian, J. (ed.) TCC 2005. LNCS, vol. 3378, pp. 325–341. Springer, Heidelberg (2005)
7. Canetti, R., Ishai, Y., Kumar, R., Reiter, M., Rubinfeld, R., Wright, R.: Selective private function evaluation with applications to private statistics. In: Proc. of the 20th Annual ACM Symposium on Principles of Distributed Computing, pp. 293–304 (2001)
8. Du, W., Zhan, Z.: Using randomized response techniques for privacy-preserving data mining. In: Proc. of the Ninth ACM SIGKDD International Conference on Knowledge Discovery and Data Mining, pp. 505–510 (2003)
9. ElGamal, T.: A public-key cryptosystem and a signature scheme based on discrete logarithms. IEEE Transactions on Information Theory IT-31(4) (1985)
10. Evfimievski, A., Gehrke, J., Srikant, R.: Limiting privacy breaches in privacy preserving data mining. In: Proc. of the 22nd ACM SIGMOD-SIGACT-SIGART Symposium on Principles of Database Systems, pp. 211–222 (2003)
11. Evfimievski, A., Srikant, R., Agrawal, R., Gehrke, J.: Privacy preserving mining of association rules. In: Proc. of the Eighth ACM SIGKDD International Conference on Knowledge Discovery and Data Mining, pp. 217–228 (2002)
12. Feigenbaum, J., Ishai, Y., Malkin, T., Nissim, K., Strauss, M., Wright, R.: Secure multiparty computation of approximations. ACM Transactions on Algorithms 2(3), 435–472 (2005)
13. Freedman, M., Nissim, K., Pinkas, B.: Efficient private matching and set intersection. In: Cachin, C., Camenisch, J.L. (eds.) EUROCRYPT 2004. LNCS, vol. 3027, pp. 1–19. Springer, Heidelberg (2004)
14. Goethals, B., Laur, S., Lipmaa, H., Mielikäinen, T.: On private scalar product computation for privacy-preserving data mining. In: Park, C.-s., Chee, S. (eds.) ICISC 2004. LNCS, vol. 3506, Springer, Heidelberg (2005)
15. Goldreich, O.: Foundations of Cryptography, Volume II: Basic Applications. Cambridge University Press, Cambridge (2004)
16. Goldreich, O., Micali, S., Wigderson, A.: How to play ANY mental game. In: Proc. of the 19th Annual ACM Conference on Theory of Computing, pp. 218–229 (1987)
17. Huang, Z., Du, W., Chen, B.: Deriving private information from randomized data. In: Proceedings of the ACM SIGMOD Conference (2005)
18. Indyk, P., Woodruff, D.: Polylogarithmic private approximations and efficient matching. In: Prof. of the Third Theory of Cryptography Conference. LNCS, Springer, Heidelberg (2006)
19. Jagannathan, G., Wright, R.N.: Privacy-preserving distributed k-means clustering over arbitrarily partitioned data. In: Proc. of the 11th ACM SIGKDD International Conference on Knowledge Discovery and Data Mining, pp. 593–599 (2005)
20. Jagannathan, G., Wright, R.N.: Privacy-preserving data imputation. In: Proc. of the ICDM Int. Workshop on Privacy Aspects of Data Mining, pp. 535–540 (2006)

21. Kantarcioglu, M., Clifton, C.: Privacy-preserving distributed mining of association rules on horizontally partitioned data. In: Proc. of the ACM SIGMOD Workshop on Research Issues on Data Mining and Knowledge Discovery (DMKD 2002), pp. 24–31 (June 2002)
22. Kantarcioglu, M., Vaidya, J.: Privacy preserving naive Bayes classifier for horizontally partitioned data. In: IEEE Workshop on Privacy Preserving Data Mining (2003)
23. Kargupta, H., Datta, S., Wang, Q., Sivakumar, K.: On the privacy preserving properties of random data perturbation techniques. In: The Third IEEE International Conference on Data Mining (2003)
24. Laur, S., Lipmaa, H., Mielikäinen, T.: Cryptographically private support vector machines. In: Proceedings of the 12th ACM SIGKDD International Conference on Knowledge Discovery and Data Mining, pp. 618–624 (2006)
25. Lindell, Y., Pinkas, B.: Privacy preserving data mining. J. Cryptology 15(3), 177–206 (2002)
26. Liu, K., Kargupta, H., Ryan, J.: Multiplicative noise, random projection, and privacy preserving data mining from distributed multi-party data. Technical Report TR-CS-03-24, Computer Science and Electrical Engineering Department, University of Maryland, Baltimore County (2003)
27. Meng, D., Sivakumar, K., Kargupta, H.: Privacy-sensitive Bayesian network parameter learning. In: Proc. of the Fourth IEEE International Conference on Data Mining, Brighton, UK (2004)
28. Rizvi, S., Haritsa, J.: Maintaining data privacy in association rule mining. In: Proc. of the 28th VLDB Conference (2002)
29. Vaidya, J., Clifton, C.: Privacy preserving association rule mining in vertically partitioned data. In: Proc. of the Eighth ACM SIGKDD International Conference on Knowledge Discovery and Data Mining, pp. 639–644 (2002)
30. Vaidya, J., Clifton, C.: Privacy-preserving k-means clustering over vertically partitioned data. In: Proc. of the Ninth ACM SIGKDD International Conference on Knowledge Discovery and Data Mining, pp. 206–215 (2003)
31. Vaidya, J., Clifton, C.: Privacy preserving naive Bayes classifier on vertically partitioned data. In: 2004 SIAM International Conference on Data Mining (2004)
32. Vaidya, J., Clifton, C.: Privacy-preserving decision trees over vertically partitioned data. In: The 19th Annual IFIP WG 11.3 Working Conference on Data and Applications Security (2005)
33. Yang, Z., Subramaniam, H., Wright, R.N.: Experimental analysis of a privacy-preserving scalar product protocol. International Journal of Computer Systems Science and Engineering 21(1), 47–52 (2006)
34. Yang, Z., Wright, R.: Privacy-preserving computation of Bayesian networks on vertically partitioned data. IEEE Transactions on Data Knowledge Engineering 18(9) (2006)
35. Yao, A.: How to generate and exchange secrets. In: Proc. of the 27th IEEE Symposium on Foundations of Computer Science, pp. 162–167 (1986)

Preserving the Privacy of Sensitive Relationships in Graph Data

Elena Zheleva and Lise Getoor

Computer Science Department
University of Maryland
College Park, MD
{elena,getoor}@cs.umd.edu

Abstract. In this paper, we focus on the problem of preserving the privacy of sensitive relationships in graph data. We refer to the problem of inferring sensitive relationships from anonymized graph data as *link re-identification*. We propose five different privacy preservation strategies, which vary in terms of the amount of data removed (and hence their utility) and the amount of privacy preserved. We assume the adversary has an accurate predictive model for links, and we show experimentally the success of different link re-identification strategies under varying structural characteristics of the data.

Keywords: privacy, anonymization, identification, link mining, social network analysis, noisy-or, graph data

1 Introduction

The goal of data mining is discovering new and useful knowledge from data. Sometimes, the data contains sensitive information, and it needs to be sanitized before it is given to data mining researchers and the public in order to address privacy concerns. Data sanitization is a complex problem in which hiding private information trades off with utility reduction. The goal of sanitization is to remove or change the attributes of the data which help an adversary infer sensitive information. The solution depends on the properties of the data and the notions of privacy and utility in the data.

Most of the work in this area makes the assumption that the data is described by a single table with attribute information for each of the entries. However, real-world datasets often exhibit more complexity. Relational data, often represented as a multi-graph, can exhibit rich dependencies between entities. The challenge of anonymizing graph data lies in understanding these dependencies and removing sensitive information which can be inferred by direct or indirect means.

Very little work has been done in this direction, and there has been a growing interest in it. The existing work looks at the identifying structural properties of the graph nodes [2,7], or considers relations to be attributes of nodes [13]. Our work assumes that the anonymized data will be useful only if it contains both structural properties and node attributes. We study anonymization techniques to match this assumption.

F. Bonchi et al. (Eds.): PinKDD 2007, LNCS 4890, pp. 153–171, 2008.

Another distinction of our approach is that, unlike existing work on privacy preservation which concentrates on hiding the identity of entities, we look at the case where relationships between entities are to be kept private. Finding out about the existence of these sensitive relationships leads to a privacy breach. We refer to the problem of inferring sensitive relationships from anonymized graph data as *link re-identification*.

Examples of sensitive relationships can be found in social networks, communication data, search engine data, disease data and others. In social network data, based on the friendship relationships of a person and the public preferences of the friends such as political affiliation, it may be possible to infer the personal preferences of the person in question as well. In cell phone communication data, finding that an unknown individual has made phone calls to a cell phone number of a known organization can compromise the identity of the unknown individual. Another example is in search data: being able to link search queries made by the same individual can give personal information that helps identify that individual. In hereditary disease data, knowing the family relationships between individuals who have been diagnosed with hereditary diseases and ones that have not, can help infer the probability of the healthy individuals to develop these diseases.

We consider the node data to be anonymized using a known single-table definition such as k-anonymization [16] or the more recently proposed t-closeness [8]. For the edge data, we propose five different anonymization strategies. The most conservative approach is to remove the relationships altogether, thus preserving any privacy that these relationships may compromise. We assume that while all of the sensitive relationships are removed, all or a portion of the relationships of other types are left intact in the anonymized data. We propose a method which allows modeling the influence of data attributes on sensitive relationships, and studying how different anonymization techniques can preserve privacy. The privacy breach is measured by counting the number of sensitive relationships that can be inferred from the anonymized data. The utility of the data is measured by counting how many attributes or observations have to be deleted in the sanitization process.

To formalize privacy preservation, Chawla et al. [4] propose a framework based on the intuitive definition that "our privacy is protected to the extent we blend in the crowd." What needs to be specified in this general framework is an abstraction of the concept of a database, the adversary information and its functionality, and when an adversary succeeds. Starting from this idea, we define the relational privacy framework for link re-identification. After the background overview in Section 2, we define the data model in Section 3. We then discuss methods for anonymizing graph data and the resulting adversary information in Section 4. Section 5 covers graph-based privacy attacks, Section 6 discusses general link re-identification attacks, and Section 7 discusses link re-identification in anonymized data and when an adversary succeeds. Section 8 presents the benefits and disadvantages of each anonymization method in an experimental setting, and Section 9 contains concluding remarks and ideas for future work.

2   Background and Related Work

Until recently, the literature on privacy preservation considered the data to be a single table, in which the rows represent records, and the columns represent attributes [1,3,8,9,12,18]. However, real-world data is often relational, and records may be related to one another or to records from other tables. For example, a database for studying hereditary diseases can contain both patient medical records and family relationships between patients. A database for studying the social network structure in a university department can contain both student information together with enrollment and research group data. Another example is data for studying Internet traffic, in which the sequences of packet traces are related to each other [14].

It is well known that even in single-table data, removing the identifying information such as social security number is not enough for preserving the privacy of individuals represented in data [18]. One of the most popular techniques for anonymizing single table data is k-anonymity, in which the quasi-identifying attributes of the table records are altered in a way that each record becomes indistinguishable from at least $k - 1$ other records [16]. The set of records with the same anonymized attributes forms an equivalence class. Since k-anonymity was first introduced, various methods for k-anonymizing data have been developed in the research community [1,3]. Recently proposed anonymity definitions such as l-diversity [9] and t-closeness [8] address some of the deficiencies of k-anonymity. l-diversity addresses the concern that an equivalence class may not contain diverse enough sensitive attributes. t-closeness addresses the stronger concern that the distribution of sensitive attributes in an equivalence class may not match the distribution of sensitive attributes in the whole data set. More definitions of privacy and information disclosure can be found in [4,5,10,11].

While it is possible to represent the nodes of a graph in a single table if the nodes have the same type, it is not clear how to do that when the nodes exhibit relationships and when there are nodes of different types. Very little work has been done on privacy preservation in graph data. Only recently, there has been privacy research on identifying structural properties of graph nodes [2,7], or on applying k-anonymity to multi-relational data [13]. The model of Miklau et al. [7] defines k-candidate anonymity for graph data based on the degrees of the nodes in the neighborhoods of the nodes to be anonymized. Their experiments on real-world datasets show that the more someone knows about the neighborhood of a node, the higher the probability for this node to be identified uniquely. They create an approach for anonymizing structure by random deletion and addition of edges. Their model assumes that the nodes and edges do not contain any attributes besides a random identifier; here, we consider models with attributes and links.

Similarly, Backstrom et al. [2] consider graphs in which the structural properties of the anonymized nodes can help an adversary to find the real-world entities behind these nodes. They consider social networks in which the node attributes are stripped off, and the edges are kept intact. They describe two families of attacks on the privacy of communication in these networks: active and passive attacks. In the active attacks, the adversary "inserts" himself in the network by creating connections with people of interest, and then tries to find himself in the anonymized version. These attacks assume that the owner of the data releases the full graph data periodically. The passive attacks assume that the adversary and his colluding friends can identify themselves in the network.

Nergiz and Clifton [13] recognize the problem that existing k-anonymizing approaches apply only to single-table data, and they extend k-anonymity to apply to relational data. Their approach abstracts the knowledge about a private entity from multiple tables into a k-anonymized tree. It keeps relationships between entities of different types but it does not discuss relationships between the entities whose privacy is a concern. Not keeping such relationships would remove some of the structural properties which are interesting in graph data.

Privacy preservation in graph data is closely related to link mining. Graph data exhibits dependencies, and they can be used to learn about identities, classes and relationships represented in it. They have been studied in the link mining community [6], and the techniques developed for collective classification, object identification and link prediction can be used to learn hidden properties of the data. If these hidden properties are sensitive, then there is a privacy breach. In this paper, we are mostly concerned with link prediction. Link prediction uses properties of the graph in order to determine whether two nodes in the graph exhibit a relationship of a particular type. For example, it may predict whether two people in a social network graph are likely to be friends. The knowledge that two people have many opportunities for communication makes them more likely to be friends, and it can be exploited by an adversary to predict likely friendships.

3   Data Model

We consider graph data which describes entities and relationships between entities. We assume that the relationships are binary relationships. In a graph, entities are represented by nodes, and relationships by edges. In general, we can have different types of nodes and different types of edges. For the purposes of this paper, we focus on the case where there is a single node type and multiple edge types. We distinguish one of the relationship types as the *sensitive relationship*. This is the relationship which we are interested to hide from the adversary. The nodes and edges can have associated attributes. In addition, the graph has structural properties. Structural properties of a node include node degree and neighborhood structure.

More formally, we consider a database describing a multi-graph $G = (V, E^1, \ldots, E^k, E^s)$, composed of a set of nodes V and sets of edges $E^1, \ldots, E^k, E^s$. Each node v_i represents an entity of interest. An edge $e^1_{i,j}$ represents a relationship of type E^1 between two nodes v_i and v_j. The $E^1, \ldots, E^k$ are the *observed* relationships, and E^s is the sensitive relationship, meaning that it is undesirable to disclose the e^s edges to the adversary.

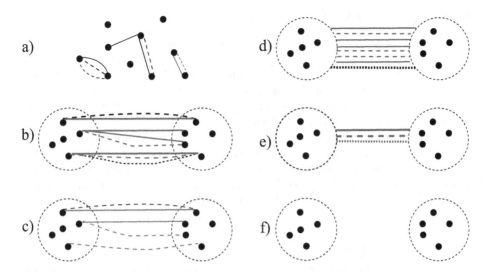

Fig. 1. The original data graph (a)) and the output from five anonymization approaches to graph data: b) revealing the observations between nodes, c) removing 50% of the observations , d) revealing all the observations between equivalence classes of nodes (cluster-edge anonymization), e) constrained revealing of the observations between equivalence classes of nodes (cluster-edge anonymization with constraints), f) removing all relational observations. There are three different edge types in the original data graph represented by different line styles. Clusters resulting from node anonymization are circled with dotted lines.

In the process of anonymizing the data, the sensitive relationships are always removed, i.e., they are not provided in the released data. However, it may be possible to predict some of these relationships using other observed relationships and/or node attributes. For the purposes of this paper, we focus on predicting sensitive edges based on the observed edges, but it is straightforward to include node and edge attributes and interesting to also consider structural properties. If the sensitive edges can be identified, then we say that there has been a *privacy breach*.

In addition, the data can include certain *constraints* which specify the number of relationships of a particular type or the number of relationships connecting any two nodes. Constraints can also be inequality constraints describing the maximum or minimum number of relationships.

As a motivating example, consider the case where the entities are students, and the relationships between students v_i and v_j include taking a class c together (`classmates`(v_i, v_j, c)), belonging to the same research group (`groupmates`(v_i, v_j, g)), and being friends (`friends`(v_i, v_j)). We can consider the class and research groups as attributes of the edges, so that students can take more than one class together, and they can belong to more than one research group. In this case, we may consider `friends` to be the sensitive relationship. We are interested in understanding how difficult it is to determine friendship based on class and research group rosters.

4   Graph Anonymization

The process of anonymization involves taking the unanonymized graph data, making some modifications, and constructing a new *released graph* which will be made available to the adversary. The modifications include changes to both the nodes and edges of the graph. We discuss several graph anonymization strategies and, for each approach, we discuss the tradeoffs between privacy preservation and the utility of the anonymized data.

We assume that the adversary has the information contained in the released graph data, and the constraints on the data. The *adversary succeeds* when she can figure out whether two nodes exhibit a sensitive relationship, i.e., when she is able to correctly predict a sensitive link between them. For example, if the adversary can figure out which students are likely to be friends given the released graph, then the data discloses private information about the two individuals.

4.1   Node Anonymization

We assume that the nodes have been anonymized with one of the techniques introduced for single table data. For example, the nodes could be k-anonymized using t-closeness [8]. This anonymization provides a clustering of the nodes into m equivalence classes $(C_1, \ldots, C_m)$ such that each node is indistinguishable in its quasi-identifying attributes from some minimum number of other nodes. We use the following notation $C(v_i) = C_k$ to specify that a node v_i belongs to equivalence class C_k.

The anonymization of nodes creates equivalent classes of nodes. Note, however, that these equivalent classes are based on node attributes only, and inside each equivalence class, there may be nodes with different identifying structural properties and edges.

4.2   Edge Anonymization

For the relational part of the graph, we describe five possible anonymization approaches. They range from one which removes the least amount of information to a very restrictive one, which removes the greatest amount of relational data. Figure 1(a) shows a simple data graph in which there are ten nodes and eight observed edges. There are three edge types, and each one is represented by a different line style. We will illustrate each of our techniques on this graph. For each approach, we discuss the tradeoffs between privacy preservation and the utility of the anonymized data.

Intact edges. The first (trivial) edge anonymization option is to only remove the sensitive edges, leaving all other observational edges intact. Figure 1(b) shows an illustration of this technique applied to the original data graph of Figure 1(a).

In our running example, we remove the friendship relationships, since they are the sensitive relationships, but we leave intact the information about students taking classes together and being members of the same research group. Since the relational observations remain in the graph, this anonymization technique should have a high utility. But it is likely to have low privacy preservation.

Intact-Edge Anonymization Algorithm ─────────────────────────────────
1: Input: $G = (V, E^1, \ldots, E^s)$
2: Output: $G' = (V', E^{1'}, \ldots, E^{k'})$
3: V'=anonymize-nodes(V)
4: **for** t=1 to k **do**
5: $\quad E^{t'} = E^t$
6: **end for**

Fig. 2. Algorithm for anonymizing graph data by removing only the sensitive edges

Partial-edge removal. Another anonymization option is to remove some portion of the relational observations. We could either remove a particular type of observation which contributes to the overall likelihood of a sensitive relationship, or remove a certain percentage of observations that meet some pre-specified criteria (e.g., at random, connecting high-degree nodes, etc.). Figure 1(c) shows an illustration of this technique when the edges are removed at random.

This partial edge removal process should increase the privacy preservation and reduce the utility of the data as compared to the previous method. Removing observations should reduce the number of node pairs with highly likely sensitive relationships but it does not remove them completely. For those pairs of nodes, private information may be disclosed.

Partial-Edge Anonymization Algorithm ─────────────────────────────
1: Input: $G = (V, E^1, \ldots, E^k, E^s)$, percent-removed
2: Output: $G' = (V', E^{1'}, \ldots, E^{k'}))$
3: V'=anonymize-nodes(V)
4: **for** t=1 to k **do**
5: $\quad E^{t'} = E^t$
6: $\quad$ removed = $\lceil$percent-removed $\times \|E^{t'}\|\rceil$
7: $\quad$ **for** i=1 to removed **do**
8: $\quad\quad e_i$ = random edge from $E^{t'}$
9: $\quad\quad E^{t'} = E^{t'} \setminus \{e_i\}$
10: $\quad$ **end for**
11: **end for**

Fig. 3. Algorithm for anonymizing graph data by removing randomly a portion of the observed edges

Cluster-edge anonymization. In the above approaches, while the nodes had been anonymized, the number of nodes in the graph was still the same, and the edges were essentially between copies of the anonymized nodes. Another approach is to collapse the anonymized nodes into a single node for each cluster, and then consider which edges to include in the collapsed graph.

The simplest approach is to leave the sets of edges intact, and maintain the counts of the number of edges between the clusters for each edge type. We refer to this technique as *cluster-edge* anonymization. Figure 4 presents the algorithm for this technique, and Figure 1(d) shows an illustration of the result from applying the algorithm.

Cluster-Edge Anonymization Algorithm ――――――――――――――――――――――――

1: Input: $G = (V, E^1, \ldots, E^k, E^s)$,
2: Output: $G' = (V', E^{1'}, \ldots, E^{k'}))$
3: $V' = \{C_1, \ldots, C_m\}$
4: **for** t=1 to k **do**
5:     $E^{t'} = \emptyset$
6:     **for all** $(v_i, v_j) \in E^t$ **do**
7:         $C_i = C(v_i)$
8:         $C_j = C(v_j)$
9:         $E^{t'} = E^{t'} \cup \{(C_i, C_j)\}$
10:    **end for**
11: **end for**

Fig. 4. Algorithm for cluster-edge anonymization technique

Cluster-edge anonymization with constraints. Next, we consider using a stricter method for sanitizing observed edges than the previous technique. The *cluster-edge anonymization with constraints* technique creates edges between equivalence classes as above, but it requires the equivalence class nodes to have the same constraints as any two nodes in the original data. For example, if there can be at most two edges of a certain type between entities, there can be at most two edges of a certain type between the cluster nodes. This, in effect, removes some of the count information that is revealed in the previous anonymization technique.

Cluster-Edge Anonymization with Constraints Algorithm ―――――――――――――――――

1: Input: $G = (V, E)$
2: Output: $G' = (V', E')$
3: $V' = \{C_1, \ldots, C_m\}$
4: **for** t=1 to k **do**
5:     $E^{t'} = \emptyset$
6:     **for all** $(v_i, v_j) \in E^t$ **do**
7:         $C_i = C(v_i)$
8:         $C_j = C(v_j)$
9:         **if** $(C_i, C_j) \notin E^{t'}$ **then**
10:            $E^{t'} = E^{t'} \cup \{(C_i, C_j)\}$
11:        **end if**
12:    **end for**
13: **end for**

Fig. 5. Algorithm for cluster-edge with constraints anonymization technique

In order to determine the number of edges of a particular type connecting two equivalence classes, the anonymization algorithm picks the maximum of the number of edges of that type between any two nodes in the original graph. In our earlier example, if the maximum number of common classes that any pair of students from the two equivalence classes takes is one class together, then the equivalence classes are connected by one class edge. Figure 1(e) shows an illustration of this technique.

This information will keep some of the utility of the data but it will say nothing of the distribution of observations. The anonymized data hides whether all observations appear on one two-node edge or on all two-node edges, and whether they ever appear in the same two-node edge. This may reduce the privacy breach on each sensitive relationship.

Removed edges. The most conservative anonymization option is to remove all the edges. Depending on the intended uses of an anonymized social network, removing the node and/or edge attributes completely may be undesirable. For example, if one wants to know whether any first-year students took a particular course together, then all the three types of information, i.e., edges, edge attributes and node attributes, are necessary. In our toy example, while taking a course together is information contained in a network edge, the name of the course is an edge attribute, and the year of enrollment is a node attribute. In this case, this anonymization technique would lead to very low utility, yet high privacy preservation.

No-Edge Anonymization Algorithm _____
1: Input: G=(V,E)
2: Output: G'=(V',∅)
3: V'=anonymize-nodes(V)

Fig. 6. Algorithm for anonymizing graph data by removing the edges

5   Graph-Based Privacy Attacks

According to Li et. al. [8], there are two types of privacy attacks in data: *identity disclosure* and *attribute disclosure*. In graph data, there is a third type of attack: *link re-identification*. Identity disclosure occurs when the adversary is able to determine the mapping from an anonymized record to a specific real-world entity (e.g. an individual). Attribute disclosure occurs when the adversary is able to infer the attributes of a real-world entity more accurately than it would be possible before the data release. Identity disclosure often leads to attribute disclosure [8]. Both identity disclosure and attribute disclosure have been studied very widely in the privacy community [1,2,3,4,7,8,9,11,12,13,16,18].

Rather than focus on these two kinds of attack, the focus of our paper is on link re-identification. Link re-identification is the problem of inferring that two entities participate in a particular type of sensitive relationship or communication. *Sensitive conclusions* are more general statements that an adversary can make about the data, and can involve both node, edge and structural information. These conclusions can be the results of aggregate queries. For example, in a database describing medical data informal about company employees, finding that almost all people who work for a particular company have a drinking problem may be undesirable. Depending on the representation of the data, this can be revealed by using both the node attributes and the co-worker relationship.

6   Link Re-identification Attacks

The extent of a privacy breach is often determined by data domain knowledge of the adversary. The domain knowledge can influence accurate inference in subtle ways. The goal of the adversary is to determine whether a sensitive relationship exists. There are different types of information that can be used to infer a sensitive relationship: node attributes, edge existence, and structural properties. Based on the domain knowledge of the adversary, she can construct rules for finding likely sensitive relationships. In this work, we assume that the adversary has an accurate probabilistic model for link prediction, which we will describe below.

In our running example, the sensitive friendship link may be re-identified based on node attributes, edge existence or structural properties. For example, consider two student nodes containing a boolean attribute "Talkative." Two nodes that both have it set to "true" may be more likely to be friends than two nodes that both have it set to "false." This inference is based on node attributes. An example of re-identification based on edge existence is two students in the same research group who are more likely to be friends compared to if they are in different research groups. A re-identification that is based on a structural property such as node degree would say that two students are more likely to be friends if they are likely to correspond to high degree nodes in the graph. A more complex observation is one which uses the result of an inferred relationship. For example, if each of two students is highly likely to be a friend with a third person based on other observations, then the two students are more likely to be friends too.

6.1   Link Re-identification Using Observations

We assume that the adversary has a probabilistic model for predicting the existence of a sensitive edge based on a set of observations O: $P(e_{ij}^s|O)$. In this work, we assume a simple *noisy-or model* [15] for the existence of the sensitive edge. The noisy-or model can capture the fact that each observed edge contributes (in a probabilistic way) to the probability of the sensitive edge existing; it makes the simplifying assumption that each factor is an independent cause for the sensitive edge. Here, we focus on re-identification based on edge existence, so the observations that we consider are sets of edges, e_{ij}^l. For simplicity, we label these observations $o_1, \ldots, o_n$. For each observed edge, we assume that we have a *noise* parameter, $\lambda_1, \ldots, \lambda_n$, and, in addition, we have a *leak* parameter λ_0 which captures the probability that the sensitive edge is there due to other, unmodeled, reasons. A noise parameter λ_i captures the independent influence of an observed relationship o_i on the existence of a sensitive relationship. Then, according to the noisy-or model, the probability of a sensitive edge is:

$$P(e^s_{ij} = 1) = P(e^s_{ij} = 1|o_1, ..., o_n) = 1 - \prod_{l=0}^{n}(1 - \lambda_l)$$

The above formula applies only when the observations are certain. It is also possible that the observation existence is not known. In that case, there are probabilities $P(o_1), \ldots, P(o_n)$ associated with the existence of each observation, and the probability of a sensitive edge is:

$$P(e^s_{ij} = 1) = \sum_{\{o\}} P(e^s_{ij} = 1|o) \prod_{k=1}^{n} P(o_k)$$

where

$$P(e^s_{ij} = 1|o) = 1 - (1 - \lambda_0) \prod_{l=1}^{n}(1 - \lambda_l)^{o_l}$$

More details about this model can be found in [17].

The noisy-or function is applicable when there are a few observations that can cause an event, and each one can contribute positively to the likelihood of the event, independent of the rest. The function has some nice properties: 1) the result of it is always between 0 and 1 when the input probabilities are in that range; 2) the final result is independent of the order in which the observations are added; 3) it can accommodate different number of observations; 4) adding a new positive observation always increases the overall likelihood. We use this function to measure how likely each sensitive relationship is, and to find whether there are parts of the graph that are vulnerable to an adversary attack. It is also possible to express the dependence between events in an explicit probability model such as a Bayesian or a Markov network, when the dependences between observations are known.

6.2   Amount of Information Disclosed

Based on the noisy-or model for each pair of nodes, it is possible to determine the number of node pairs that are likely to participate in a sensitive relationship. In the anonymized data, it is desirable to have few sensitive relationships which can be inferred with high likelihood. To formalize this desirable property, we can compute the percentage of all possible two-node relationships which have a high likelihood and make sure that it is below some allowed level δ:

$$\frac{|relationships(P(e^s_{ij}) > \rho)|}{|V|^2} < \delta \tag{1}$$

where ρ is the threshold for predicting that a sensitive relationship exists and $relationships(P(e^s_{ij}) > \rho)$ returns the set of all sensitive relationships which have likelihood above ρ. For example, if it is true for the given data that 15% of the possible pair relationships have a true likelihood of exhibiting a sensitive relationship higher than 0.8, then

$$\frac{|relationships(P(e_{ij}^s) > 0.8)|}{|V|^2} <= 0.15.$$

For each anonymization technique, it is possible to find the highest possible δ that satisfies a particular ρ level. This can be used to compare the privacy preservation for each technique. The higher the δ, the lower the privacy preservation.

6.3   Utility

Utility in the data is hard to measure, and we make an assumption that the more observations there are in the anonymized data, the better. To measure utility, we use a very simple approach. We count the number of observations which were removed in the process of anonymization. The lower the number of removed observations, the higher the overall utility. For the intact edge and the cluster-edge anonymization techniques, no relational observations are deleted, therefore, these two techniques have the highest utility. For the partial edge removal technique, the utility depends on the percentage of edges removed. For the cluster-based with constraints technique, it is much lower, since the graph is collapsed, and many edges are removed. The exact number can be computed using the properties and constraints of the data such as number of nodes, edges of each type, and the size of the equivalence classes. Note that a more sophisticated measure of utility would also consider the loss of structural properties in the anonymized data. In the case when all the edges are removed, the utility is 0.

7   Link Re-identification in Anonymized Data

In the first two types of link anonymization (intact and partial), the noisy-or model can be used directly to compute the probability of a sensitive edge. In the other two cases, one has to consider the probability that an observed edge exists between two nodes, and apply the noisy-or.

7.1   Link Re-identification in Cluster-Edge Anonymization

In the case of keeping edges between equivalence classes, the probability of an observation existing between two nodes is not given and it needs to be estimated. The noisy-or function will need to take into consideration the probability associated with each observation in order to compute the likelihood of a sensitive relationship. When the number of relationships of each type (e.g., course, research group, etc.) between two equivalence classes is given, the distribution is not uniform, and the probability of an observation $P(o) = P(observation(v_i, v_j))$ existing between two students can be computed directly from the counts of relationships between their equivalence classes. $P(\text{classmates}(v_i, v_j, c))$ expresses the probability that there exists a class edge between any two students v_i and v_j from two equivalence classes $C(v_i)$ and $C(v_j)$, i.e., the students take a course c together. It is equal to the number of possible student pairs from the two equivalence classes who take a course together —$\text{classmates}(C(v_i), C(v_j))$— as a fraction of the number of possible relationships in the graph $|V|^2$.

7.2 Link Re-identification in Cluster-Edge Anonymization with Constraints

In the constrained cluster-edge anonymization approach, the number of relationships between equivalence classes is not given. Therefore, the probability of an observation existing between any two edges has to be taken into account in the noisy-or model. To estimate this probability, an adversary can assume a uniform distribution, meaning that the probability of an observation existing between any two edges is the same for all edges in the graph. This estimate is worse than the cluster-edge anonymization method. Using the constraints on the data, it is possible to get estimates of this probability. For example, if it is known that there are 50 pairs of students who take courses together, and there are 100 possible pairs, then the probability of any two students taking any class c together is $P(\texttt{classmates}(v_i, v_j, c))=0.5$. If the adversary knows the number of offered courses c, the number of courses per person n, the number of students $s = |V|$, and assumes that all courses have the same number of people $p = \frac{s*n}{c}$, then the number of possible pairs who take courses together can be calculated as $n * (p - 1)$. This number can be used to compute in a manner similar to the cluster-edge anonymization method $P(\texttt{classmates}(v_i, v_j, c))=\frac{n*(p-1)}{|V|^2}$.

One can also use an expected value of any two-node relationship to be sensitive by looking at the likelihood distribution of all relationships. However, we found that this does not measure privacy well because an adversary is more interested in the highly likely relationships.

An observation probability shows the percentage of edges between two nodes from two different equivalence classes that contain the observation. For example, if the two equivalence classes have exactly 10 nodes each, and the observation exists for 30 of the two-node edges, then the edge probability is $P(\texttt{observation}(v_i, v_j))=0.3$ where $\texttt{observation}(v_i, v_j)$ is either $\texttt{classmates}(v_i, v_j, c)$, or $\texttt{groupmates}(v_i, v_j, g)$ for any c and g. This increases the utility of the data as compared to the case when no probabilities are included, but it can also decrease the privacy preservation. An exception is the case when observations between equivalence classes have exactly the same distribution as the overall uniform distribution.

8 Experiments

The effectiveness of the anonymization approaches depends on the structural and statistical characteristics of the underlying graph. In order to study the influence of each anonymization approach on privacy preservation, we apply them to synthetic data generated under varying statistical and structural assumptions and compute the information disclosed. We show how many relationships are revealed at different probability thresholds. First, we describe the data generator.

8.1 Data Generator

The data generator creates data according to the data model described in Section 3. The input to the data generator includes: the number of nodes, maximum number of nodes which can participate in a relationship (e.g., the maximum number of students taking the same class), the maximum number of relationships that each student can have with any other student (e.g., maximum number of classes that a student can take). For all observation types, the probability of two nodes exhibiting a sensitive relationship given the observation type is given and the leak probability, the probability of two nodes exhibiting a sensitive relationship due to unobserved causes.

For the concrete example, the data generator starts by creating a set of students, a set of classes, and a set of research groups. There are constraints on how many classes each student takes, and on how many research groups each student belongs. There are also constraints on the maximum number of students per class and on the maximum number of students per group. For each student, the generator picks random classes to enroll into up to the maximum number of classes per student possible. Similarly, each student is assigned to a random research group.

The nodes in the data graph represent students. There is a `classmates` edge connecting two students for each class they take together, and there is `groupmates` edge if they belong to the same research group. These pieces of information represent observations indicating that two students may be friends, i.e., that they may exhibit a sensitive relationship. The ground truth is generated by computing the probability of a friendship between each two students using the noisy-or model, and assigning the friendship a true value with a probability equal to that likelihood.

The parameters given to the data generator can be varied. We would like to explore graphs which vary in their density, therefore we allow the number of classes and research groups to vary while fixing the number of nodes/students to 100. The constraints on the data are that each student takes two classes, and belongs to one research group. Also, a class can have no more than 25 people, and a group can have no more than 15. We picked probabilities which make sense in the domain. The prior probability of two students knowing each other is $P(\texttt{friends}(v_i, v_j))=0.2$. It is relatively high because the students are from the same department. The probability that two students know each other if they are in the same class c is $P(\texttt{friends}(v_i, v_j) - \texttt{classmates}(v_i, v_j, c))=0.4$. The probability that two students know each other if they are in the same research group is $P(\texttt{friends}(v_i, v_j) - \texttt{groupmates}(v_i, v_j, c))=0.6$.

8.2 Evaluating Privacy Preservation in Anonymized Data

We begin by studying the privacy preservation in the data that results from each of the anonymization techniques. In particular, we study the number of correctly identified sensitive relationships for the following anonymization functions: 1) when the anonymization function leaves the edges between nodes intact (4.2),

2) when it removes 50% of the observations chosen at random (4.2), 3) when it leaves edges between node equivalence classes in the cluster-edge anonymization (4.2), and 4) when it leaves edges between node equivalence classes with a constrained number of observations (4.2). For the last two, each node is assigned randomly to an equivalence class. We vary k, the number of nodes in each equivalence class, and show the results for $k = 2$ and $k = 6$ because they exhibit the tendencies of varying k well.

Correctly Identified Senstive Relationships after Anonymization

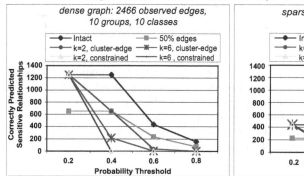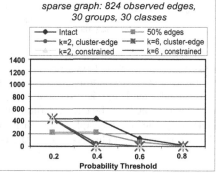

Fig. 7. Comparison between the number of sensitive relationships found after each of six anonymization techniques has been applied. The number of revealed friendships decreases as the friendship likelihood threshold increases. The two constrained cluster-edge methods (at $k = 2$ and $k = 6$) reveal the same number of relationships in both graphs. In the sparse graph, the cluster-edge method at $k = 6$ (not constrained) also overlaps with the two constrained methods.

The data was generated with the default parameters, varying the number of classes and the number of research groups between 10 and 30. A graph, in which there are 10 research groups and 10 classes, is very dense, and a graph at the other extreme with 30 research groups and 30 classes is very sparse. We show these "extreme" cases in Figure 7 and Figure 8. To account for the randomness in the generated graph, we ran the experiments on 100 generated graphs, and present the average performance. Note that when using the default data parameters (at most two classes taken by each student and at most one group of which a student is a member), the maximum possible likelihood for their friendship is 0.89.

We measure the precision, recall rate and the number of inferred sensitive relationships in the anonymized graphs. The precision shows how many of the predicted sensitive relationships are true sensitive relationships. The recall rate measures what portion of the true sensitive relationships can be predicted. Translated into the privacy domain, the recall rate measures what portion of the true sensitive relationships have been compromised, and the precision shows what is the chance that a predicted relationship is really a sensitive one. For example, if the analysis predicts 10 sensitive relationships and only 5 of them are true, then the precision is 0.5. If there are a total of 100 true sensitive relationships in the

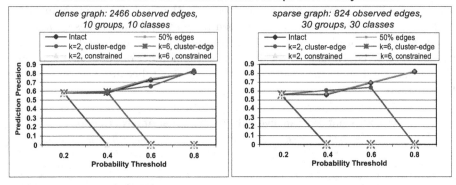

Fig. 8. Comparison between the precision of predicted friendships found after one of six anonymization techniques has been applied. At low threshold values, the number of revealed friendships is large but the precision is low. The precision of the method that removes 50% of the edges at random overlaps with the precision of the intact-edge method in the sparse graph, and nearly overlaps in the dense graph. The precision of the two constrained cluster-edge methods (at $k = 2$ and $k = 6$) overlap as well.

network, then the recall rate is 0.05. Ideally, a model for predicting sensitive information would should have a high precision and a high recall rate when tested on the original data, and a low precision and a low recall rate when tested on the anonymized data.

A low precision in the anonymized data is more crucial than a low recall rate. A combination of a high precision with a low recall rate in the anonymized data is undesirable because it means that the anonymization can hide most of the sensitive relationships but the ones that can be predicted are highly likely to be true. Results with a low precision and a high recall rate are not as bad. In this case, even though the anonymization allows many of the true sensitive relationships to be predicted, the true sensitive relationships are indistinguishable from many non-sensitive relationships.

8.3   Results

Figure 7 shows a comparison between the number of sensitive relationships inferred after each of six anonymization techniques has been applied. It shows that at higher thresholds (0.6 and 0.8), keeping all the edges between node equivalence classes preserves privacy much better than deleting 50% of the two-node edges, while having higher utility as discussed in Section 6.3. As expected, for lower k, the privacy preservation is lower: the number of revealed relationships is higher in the data anonymized with the cluster-edge method. In the data anonymized with the cluster-edge method with constraints, varying k yielded to the same results, which is why the graphs of $k = 2$ overlap with the graphs, in which $k = 6$.

Prediction Precision and Recall Rates at Various Classmate Densities

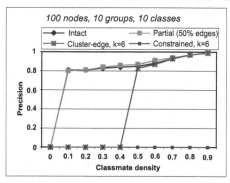
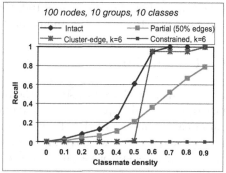

Fig. 9. Comparison between the precision at different classmate density levels (a)) shows that at high density levels, the cluster-edge anonymization preserve privacy as badly as the anonymization which deletes 50 % of the edges. Moreover, the recall rate at these levels (b)) is much higher for the cluster-edge method. The groupmate density is kept constant at 0.1.

We also ran the experiments for other combinations of class and group parameters in the range [10,30]. The experiments confirmed that as the number of observed edges decreases, so does the number of correctly identified sensitive relationships. However, the behavior at different thresholds is proportionately the same for all anonymization methods except the cluster-edge method. In the cluster-edge method, the privacy is preserved better in the sparse graph for both k levels, as seen by comparing the dense and the sparse graph results at threshold 0.4. In the sparse graph, the results when $k = 6$ are the same as the ones of the cluster-edge with constraints.

Figure 8 shows that even though lower probability thresholds reveal more sensitive relationships, the precision is low. At higher probability thresholds, the precision is high but on a very small number of predicted relationships.

Experimenting with the number of nodes in the network showed that the precision and sensitivity results were invariant to the network size when the friendship, groupmate and classmate densities were kept constant. The density values were 0.36, 0.1 and 0.2, respectively. The tested networks were of size 100, 200, 300 and 400 nodes. Other constant parameters were the number of groups, 10, the number of classes, 10, and the k-anonymization parameter $k = 6$.

We also varied the multigraph classmate density by varying the number of classes each student joined. Since this parameter was used in the data generator as well, it affected the friendship density of the original graph. The correlation between the two densities was positive. We found that at high classmate density levels the claim that the cluster-edge anonymization preserves privacy better than the anonymization which deletes 50% of the edges no longer held. As Figure 9a) shows that as the class density goes above 0.4 (friendship density is 0.63), the precision of predicted sensitive links is almost the same for the two methods. Moreover, as Figure 9b) at levels above 0.5 (friendship density is 0.76), the data anonymized with the cluster-edge method has much higher recall rate. Again, the number of nodes was 100, the number of groups was 10, the number of classes was 10, and the k-anonymization parameter k was 6.

9   Conclusion

In this paper, we have focused on the problem of link re-identification. We have proposed several approaches for anonymizing graph data and done an initial empirical evaluation of the effectiveness of the different strategies. The work is preliminary, in that we have made very specific assumptions about the model and the data generator parameters. However, because understanding and appreciating the subtleties in the effectiveness of techniques is such an important and timely topic, we hope that this work will motivate further research in the topic.

References

1. Aggarwal, G., Feder, T., Kenthapadi, K., Motwani, R., Panigrahy, R., Thomas, D., Zhu, A.: Approximation algorithms for k-anonimity. Journal of Privacy Technology (November 2005)
2. Backstrom, L., Dwork, C., Kleinberg, J.: Wherefore art thou r3579x: Anonymized social networks, hidden patterns, and structural steganography. In: 16th International Conference on World Wide Web (WWW), pp. 181–190 (2007)
3. Bayardo, R., Agrawal, R.: Data privacy through optimal k-anonymization. In: IEEE 21st International Conference on Data Engineering (April 2005)
4. Chawla, S., Dwork, C., Mcsherry, F., Smith, A., Wee, H.: Toward privacy in public databases. In: Proceedings of the Theory of Cryptography Conference (2005)
5. Evfimievski, A., Gehrke, J., Srikant., R.: Limiting privacy breaches in privacy preserving data mining. In: ACM Principles of database systems (PODS), pp. 211–222 (June 2003)
6. Getoor, L., Diehl, C.P.: Link mining: A survey. SIGKDD Explor. Newsl. 7(2), 3–12 (2005)
7. Hay, M., Miklau, G., Jensen, D., Weis, P., Srivastava, S.: Anonymizing social networks (March 2007)
8. Li, N., Li, T., Venkatasubramanian, S.: t-closeness: Privacy beyond k-anonymity and l-diversity. In: IEEE 23rd International Conference on Data Engineering, pp. 106–115 (April 2007)
9. Machanavajjhala, A., Gehrke, J., Kifer, D., Venkitasubramaniam, M.: l-diversity: Privacy beyond k-anonymity. In: 22nd IEEE International Conference on Data Engineering (2006)
10. Miklau, G., Suciu, D.: A formal analysis of information disclosure in data exchange. In: ACM Conference on Management of Data (SIGMOD), pp. 575–586 (2004)
11. Nergiz, M.E., Atzori, M., Clifton, C.: Hiding the presence of individuals from shared databases. In: 26th ACM SIGMOD International Conference on Management of Data (June 2007)
12. Nergiz, M.E., Clifton, C.: Thoughts on k-anonymization. In: IEEE 22nd International Conference on Data Engineering Workshops (ICDEW), p. 96 (April 2006)
13. Nergiz, M.E., Clifton, C.: Multirelational k-anonymity. In: IEEE 23rd International Conference on Data Engineering Posters (April 2007)

14. Pang, R., Paxson, V.: A high-level programming environment for packet trace anonymization and transformation. In: ACM SIGSOMM (2003)
15. Pearl, J.: Probabilistic Reasoning in Intelligent Systems: Networks of Plausible Inference, Inc., San Mateo, California. Morgan Kaufmann Publishers, San Francisco (1988)
16. Samarati, P.: Protecting respondents' identities in microdata release. Knowledge and Data Engineering 13(6), 1010–1027 (2001)
17. Singliar, T., Hauskrecht, M.: Noisy-or component analysis and its application to link analysis. Journal of Machine Learning Research 7, 2189–2213 (2006)
18. Sweeney, L.: Achieving k-anonymity privacy protection using generalization and suppression. International Journal of Uncertainty 10(5), 571–588 (2002)

Author Index

Lecture Notes in Computer Science

Sublibrary 4: Security and Cryptology

For information about Vols. 1– 3856
please contact your bookseller or Springer

Vol. 4437: J.L. Camenisch, C.S. Collberg, N.F. Johnson, P. Sallee (Eds.), Information Hiding. VIII, 389 pages. 2007.

Vol. 4392: S.P. Vadhan (Ed.), Theory of Cryptography. XI, 595 pages. 2007.

Vol. 4377: M. Abe (Ed.), Topics in Cryptology – CT-RSA 2007. XI, 403 pages. 2006.

Vol. 4356: E. Biham, A.M. Youssef (Eds.), Selected Areas in Cryptography. XI, 395 pages. 2007.

Vol. 4341: P.Q. Nguyên (Ed.), Progress in Cryptology - VIETCRYPT 2006. XI, 385 pages. 2006.

Vol. 4332: A. Bagchi, V. Atluri (Eds.), Information Systems Security. XV, 382 pages. 2006.

Vol. 4329: R. Barua, T. Lange (Eds.), Progress in Cryptology - INDOCRYPT 2006. X, 454 pages. 2006.

Vol. 4318: H. Lipmaa, M. Yung, D. Lin (Eds.), Information Security and Cryptology. XI, 305 pages. 2006.

Vol. 4307: P. Ning, S. Qing, N. Li (Eds.), Information and Communications Security. XIV, 558 pages. 2006.

Vol. 4301: D. Pointcheval, Y. Mu, K. Chen (Eds.), Cryptology and Network Security. XIII, 381 pages. 2006.

Vol. 4300: Y.Q. Shi (Ed.), Transactions on Data Hiding and Multimedia Security I. IX, 139 pages. 2006.

Vol. 4298: J.K. Lee, O. Yi, M. Yung (Eds.), Information Security Applications. XIV, 406 pages. 2007.

Vol. 4296: M.S. Rhee, B. Lee (Eds.), Information Security and Cryptology – ICISC 2006. XIII, 358 pages. 2006.

Vol. 4284: X. Lai, K. Chen (Eds.), Advances in Cryptology – ASIACRYPT 2006. XIV, 468 pages. 2006.

Vol. 4283: Y.Q. Shi, B. Jeon (Eds.), Digital Watermarking. XII, 474 pages. 2006.

Vol. 4266: H. Yoshiura, K. Sakurai, K. Rannenberg, Y. Murayama, S.-i. Kawamura (Eds.), Advances in Information and Computer Security. XIII, 438 pages. 2006.

Vol. 4258: G. Danezis, P. Golle (Eds.), Privacy Enhancing Technologies. VIII, 431 pages. 2006.

Vol. 4249: L. Goubin, M. Matsui (Eds.), Cryptographic Hardware and Embedded Systems - CHES 2006. XII, 462 pages. 2006.

Vol. 4237: H. Leitold, E.P. Markatos (Eds.), Communications and Multimedia Security. XII, 253 pages. 2006.

Vol. 4236: L. Breveglieri, I. Koren, D. Naccache, J.-P. Seifert (Eds.), Fault Diagnosis and Tolerance in Cryptography. XIII, 253 pages. 2006.

Vol. 4219: D. Zamboni, C. Krügel (Eds.), Recent Advances in Intrusion Detection. XII, 331 pages. 2006.

Vol. 4189: D. Gollmann, J. Meier, A. Sabelfeld (Eds.), Computer Security – ESORICS 2006. XI, 548 pages. 2006.

Vol. 4176: S.K. Katsikas, J. López, M. Backes, S. Gritzalis, B. Preneel (Eds.), Information Security. XIV, 548 pages. 2006.

Vol. 4117: C. Dwork (Ed.), Advances in Cryptology - CRYPTO 2006. XIII, 621 pages. 2006.

Vol. 4116: R. De Prisco, M. Yung (Eds.), Security and Cryptography for Networks. XI, 366 pages. 2006.

Vol. 4107: G. Di Crescenzo, A. Rubin (Eds.), Financial Cryptography and Data Security. XI, 327 pages. 2006.

Vol. 4083: S. Fischer-Hübner, S. Furnell, C. Lambrinoudakis (Eds.), Trust and Privacy in Digital Business. XIII, 243 pages. 2006.

Vol. 4064: R. Büschkes, P. Laskov (Eds.), Detection of Intrusions and Malware & Vulnerability Assessment. X, 195 pages. 2006.

Vol. 4058: L.M. Batten, R. Safavi-Naini (Eds.), Information Security and Privacy. XII, 446 pages. 2006.

Vol. 4047: M.J.B. Robshaw (Ed.), Fast Software Encryption. XI, 434 pages. 2006.

Vol. 4043: A.S. Atzeni, A. Lioy (Eds.), Public Key Infrastructure. XI, 261 pages. 2006.

Vol. 4004: S. Vaudenay (Ed.), Advances in Cryptology - EUROCRYPT 2006. XIV, 613 pages. 2006.

Vol. 3995: G. Müller (Ed.), Emerging Trends in Information and Communication Security. XX, 524 pages. 2006.

Vol. 3989: J. Zhou, M. Yung, F. Bao (Eds.), Applied Cryptography and Network Security. XIV, 488 pages. 2006.

Vol. 3969: Ø. Ytrehus (Ed.), Coding and Cryptography. XI, 443 pages. 2006.

Vol. 3958: M. Yung, Y. Dodis, A. Kiayias, T.G. Malkin (Eds.), Public Key Cryptography - PKC 2006. XIV, 543 pages. 2006.

Vol. 3957: B. Christianson, B. Crispo, J.A. Malcolm, M. Roe (Eds.), Security Protocols. IX, 325 pages. 2006.

Vol. 3956: G. Barthe, B. Grégoire, M. Huisman, J.-L. Lanet (Eds.), Construction and Analysis of Safe, Secure, and Interoperable Smart Devices. IX, 175 pages. 2006.

Vol. 3935: D.H. Won, S. Kim (Eds.), Information Security and Cryptology - ICISC 2005. XIV, 458 pages. 2006.

Vol. 3934: J.A. Clark, R.F. Paige, F.A.C. Polack, P.J. Brooke (Eds.), Security in Pervasive Computing. X, 243 pages. 2006.

Vol. 3928: J. Domingo-Ferrer, J. Posegga, D. Schreckling (Eds.), Smart Card Research and Advanced Applications. XI, 359 pages. 2006.

Vol. 3919: R. Safavi-Naini, M. Yung (Eds.), Digital Rights Management. XI, 357 pages. 2006.

Vol. 3903: K. Chen, R. Deng, X. Lai, J. Zhou (Eds.), Information Security Practice and Experience. XIV, 392 pages. 2006.

Vol. 3897: B. Preneel, S. Tavares (Eds.), Selected Areas in Cryptography. XI, 371 pages. 2006.

Vol. 3876: S. Halevi, T. Rabin (Eds.), Theory of Cryptography. XI, 617 pages. 2006.

Vol. 3866: T. Dimitrakos, F. Martinelli, P.Y.A. Ryan, S. Schneider (Eds.), Formal Aspects in Security and Trust. X, 259 pages. 2006.

Vol. 3860: D. Pointcheval (Ed.), Topics in Cryptology – CT-RSA 2006. XI, 365 pages. 2006.

Vol. 3858: A. Valdes, D. Zamboni (Eds.), Recent Advances in Intrusion Detection. X, 351 pages. 2006.